Electronics 3

Heinemann:London

Electronics 3

S. A. Knight B.Sc (Hons)

Lately Senior Lecturer in Mathematics & Electronic
Engineering, Bedford College of Higher Education

Heinemann Newnes
An imprint of Heinemann Professional Publishing Ltd
Halley Court, Jordan Hill, Oxford OX2 8EJ

OXFORD LONDON MELBOURNE AUCKLAND SINGAPORE
IBADAN NAIROBI GABORONE KINGSTON

First published as *Electronics for Technicians 3* by Butterworth 1980
Second edition published by Heinemann Newnes 1989

British Library Cataloguing in Publication Data
Knight, S. A. (Stephen Alfred)
    Electronics 3. – 2nd ed.
    1. Electronic equipment. For technicians
    I. Title   II. Knight, S. A. (Stephen Alfred).
    Electronics for technicians 3. 621.381

    ISBN 0–434–91079–1

Typeset by TecSet Ltd.
Printed and bound in Great Britain by
Butler & Tanner Ltd, Frome and London

Contents

Preface

There have been a considerable number of changes to the BTEC guide syllabus for Electronics 3, unit number U86/332, since this book was originally published, and a complete overhaul of its contents has been made necessary. Only three of the original sections are retained: those on the principles of negative feedback, noise and, with some modification, oscillators. The field-effect transistor and stabilized power supplies, originally in *Electronics 3*, have been earlier transferred to *Electronics 2*, where those parts of the syllabus are now to be found.

The emphasis in this book, therefore, has been on those parts of the syllabus which are often considered to be the 'poor relations' of the subject matter and which receive only a superficial coverage in many textbooks outside specialist works. These sections are decibel measurement, display devices, fibre optics, operational amplifiers, digital-to-analogue and analogue-to-digital converters, solid-state relays and controlled rectification, the unijunction transistor and fault-finding methods. Those other important parts of the syllabus, combinational and sequential logic, have been covered (as pointed out in its preface) in *Electronics 2* by extensions beyond the requirements of level II, and with the addition there of a section on Karnaugh mapping.

It is believed therefore that, between them, *Electronics 2* and *Electronics 3* in this series cover most of the pertinent work relating to the BTEC guide syllabuses for these two levels, and hopefully will enable students and lecturers to build a sufficiently wide base on which to erect their coursework.

A number of practical experiments are described throughout the text and these can, with little difficulty, be adapted and extended to provide a good range of essential laboratory work and investigative practice. Devices tend to change rapidly these days, but there should be no problems in obtaining the semiconductors and integrated circuits suggested, which have been reliably available for a number of years.

As with the other books in this series, there are plenty of worked examples. Self-assessment problems are provided throughout the text and at the end of each section, and solutions are given at the end of the book.

S. A. Knight
Market Harborough

1 Decibels

Aims: At the end of this unit section you should be able to:
Understand why the use of decibel notation is advantageous.
Perform calculations with decibel gains and losses.
Use a reference power level and explain its advantages.

THE USE OF DECIBELS

There are a number of advantages to be had by expressing power, voltage and current ratios in logarithmic units.

Consider a circuit network connecting a signal source to an output load. If the input power to the network is P_i and the output power is P_o, then the ratio of P_o to P_i or the power gain of the system is P_o/P_i. This ratio may or may not be greater than unity.

If a number of such networks are connected in cascade (which means that the output of each provides the input to the next following) as shown in *Figure 1.1*, each network introduces an

Figure 1.1

input-to-output power ratio P_2/P_i, P_3/P_2 and so on. The overall or total power gain is the product of the individual ratios, since

$$\frac{P_2}{P_i} \times \frac{P_3}{P_2} \times \frac{P_4}{P_3} \times \frac{P_5}{P_4} \times \frac{P_o}{P_5} = \frac{P_o}{P_i}$$

everything else conveniently cancelling out.

But with practical figures instead of symbols, such convenient cancellations rarely turn up and we find ourselves faced with a tedious calculation. In *Figure 1.2* six networks all having

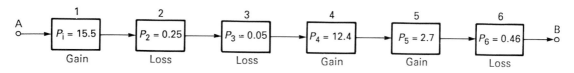

Figure 1.2

different power ratios are wired in cascade between points A and B. The individual power ratios are seen to be, in order:

$$P_1 = 15.50 \quad \text{a power gain}$$
$$P_2 = 0.25 \quad \text{a power loss}$$
$$P_3 = 0.05 \quad \text{a power loss}$$
$$P_4 = 12.40 \quad \text{a power gain}$$
$$P_5 = 2.70 \quad \text{a power gain}$$
$$P_6 = 0.46 \quad \text{a power loss}$$

The total power ratio over the system can be calculated to be

$$15.50 \times 0.25 \times 0.05 \times 12.40 \times 2.70 \times 0.46 = 2.984$$

which is an overall power gain, being greater than unity.

It often happens that the figures obtained in such calculations become large and clumsy, for example when an amplifier shows large gain variations against input frequency and when these changes need to be plotted as a graph. The problem will become clear if a range of gain ratios between, say, 10 and 1 000 000 are attempted on a piece of ordinary linear graph paper.

The use of logarithmic units enables the overall gain of a system to be calculated by addition (or subtraction) rather than the more laborious multiplication illustrated above, and additionally brings the large gain variations mentioned down to manageable size.

We know that

$$\log (P_1 \times P_2) = \log P_1 + \log P_2$$

and that

$$\log \frac{P_1}{P_2} = \log P_1 - \log P_2$$

So the ratio of two powers, P_2 to P_1 say, can be described by finding the logarithm of that ratio. This logarithmic ratio is measured in terms of a unit called the bel, and so a power ratio is given in the form

$$\log \frac{P_2}{P_1} \text{ bels}$$

The logarithms are to base 10.

In practice a unit of one bel is inconveniently large and a one-tenth part of it, the decibel, is generally used. Using this smaller unit we have a power ratio expressed as

$$10 \log \frac{P_2}{P_1} \text{ decibels (dB)}$$

With these units, if the signal power delivered to the load at the receiving end of a cable is $^1/_{10}$ that at the sending end, the *loss* is 10 dB. With two similar cables in series, the received power would be $^1/_{100}$ that transmitted and the loss would be 20 dB.

In any amplifier, if the output power is 100 times the input power, the gain is 20 dB. With two such amplifiers in cascade, the gain in power ratio would be $100 \times 100 = 10\ 000$ or 40 dB. With modern amplifiers having gains of a million or more, the use of decibels enables figures conveniently small to be used instead of those astronomically large.

Example 1
Express as decibel gains or losses the power ratios (a) 200 (b) 40 (c) 0.005.

(a) gain $= 10 \log 200$
$= 10 \times 2.301 \quad = 23.01$ dB

(b) gain = 10 log 40
  = 10 × 1.602 = 16.02 dB
(c) gain = 10 log 0.005
  = 10 × −2.301 = −23.01 dB

These solutions would normally be rounded off to 23 dB, 16 dB and −23 dB respectively.

The last case is a power loss, indicated by the negative sign. Since

$$10 \log \frac{P_2}{P_1} = -10 \log \frac{P_1}{P_2}$$

the numerical answer obtained will be the same irrespective of whether we consider the ratio as P_2/P_1 or P_1/P_2. This is illustrated by parts (a) and (c) of the example. The correct sign to indicate a power gain (+) or a power loss (−) must, however, be obtained from a consideration of whether the ratio P_2/P_1 is greater than unity or less than unity, respectively.

We can save ourselves a lot of work if we take the ratio of powers always to be greater than unity, inverting if it is less. By this device we always get a wholly positive characteristic to the logarithm and so avoid the difficulty of having to cope with negative numbers; calculators give logarithms as wholly negative where appropriate, but tables do not. Whether the answer is a gain or a loss is at once apparent by the original form of the power ratio.

Example 2
A network has an output equal to $^1/_4$ of the input. Find the loss in dB.

Let us work this problem in two ways. First we use the 'fractional' way:

gain = 10 log 0.25
  = 10 × 1.3980
  = 10 × (−1 + 0.3980)
  = 10 × −0.602 = −6.02 dB

Now to avoid the negative characteristic and the necessity of having to express the logarithm of 0.25 as a wholly negative number, let us invert the power ratio concerned and express it as $^1/_{0.25}$ or 4. Then

gain = 10 log 4
  = 10 × 0.602 dB

But the original power ratio tells us that the network is plainly introducing a power loss. Hence

loss = −6.02 dB   or   6.02 dB down

Example 3
An amplifier has a power gain of 8 dB. What is the ratio of output to input power?

Here

$$10 \log (\text{ratio}) = 8$$
$$\log (\text{ratio}) = 0.8$$
$$\text{ratio} = 10^{0.8} \quad \text{(or antilog 0.8)}$$
$$= 6.31$$

It is clear from this example that the dB is a unit of power *ratio* and not a unit of *absolute* power in any shape or form. To say that the gain of a system is 8 dB is simply to imply that the output power is 6.31 times the input power. We are told nothing about the magnitudes of the input and output powers in absolute terms.

A meaning can only definitely be attached to power ratios when we have a starting point or reference power level. If we are told, for example, that the price of a video recorder is £50 more than the price of a television receiver, we are none the wiser about the selling price of either the recorder or the televisor. The statement of 'so many dB' gain or loss has to be referred to a particular reference level. A power of so many dB above or below that level then acquires a definite meaning in quantitative terms. Such a reference level might refer to the input power of a circuit system. For an input power level of, say, 10 mW, a 6 dB gain tells us that the output power will be 40 mW. Provided we stuck to a 10 mW input level as reference, the output of any system could be unambiguously stated in dB of power.

It is a general communications practice, however, to use a reference power of 1 mW (0.001 W). Then any power gain P can be expressed as $10 \log [P/1 \text{ mW}]$ dB referred to 1 mW. Thus a power gain of 23 dB (see Example 1) representing a power gain of 200 times will, referred to 1 mW as input, represent an output of 200 mW. In the same way, a power loss of 23 dB (-23 dB) referred to 1 mW as input will represent an output power of $^1/_{200}$ or 0.005 mW. The expression 'dB referred to 1 mW' is generally abbreviated to dBm. In this way

$$20 \text{ dBm} = 20 \text{ dB with respect to } 1 \text{ mW}$$
$$= 20 \text{ dB above } 1 \text{ mW}$$
$$= 100 \text{ mW}$$

Example 4
What power, in watts, is represented by 15 dBm?

If the required power is P watts, then

$$\frac{P}{1 \text{ mW}} = 10^{1.5} \text{ (or antilog 1.5)}$$

$$= 31.63$$

Therefore

$$P = 31.63 \text{ mW} \quad \text{or} \quad 0.0316 \text{ W}$$

Values in dBm are usually rounded off to the nearest whole numbers, and *Table 1.1* gives a list of powers expressed in this way. It is useful to memorize the power *ratios* corresponding to 3 dB, 6 dB, 10 dB and 20 dB.

Table 1.1 Power decibels referred to 1 mW

μW	dBm	mW	dBm
1	−30	1	0
2	−27	2	+3
4	−24	4	+6
5	−23	5	+7
8	−21	8	+9
10	−20	10	+10
20	−17	20	+13
40	−14	40	+16
50	−13	50	+17
100	−10	100	+20
200	−7	200	+23
400	−4	400	+26
500	−3	500	+27
1000	0	1000	+30
		10W	+40

The table shows the usefulness of decibel notation in that the product of powers becomes additive. For example, 8 mW = 9 dBm and 10 mW = 10 dBm. The total power gain from two such systems in cascade is 8 × 10 = 80 mW, which from the table is equivalent to 19 dBm; and 9 dBm + 10 dBm = 19 dBm.

Powers expressed in dBm and power ratios in dB are therefore *added* (algebraically) to provide the overall gain of the system.

Example 5
Six networks are wired in cascade as shown in *Figure 1.3*, the decibel gains and losses being as indicated for each network. Find the power output if 1 mW is applied at the input.

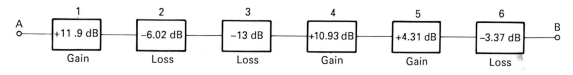

Figure 1.3

> The input in this example is 0 dBm, the reference level. The total gain is the algebraic sum of the individual gains, which is
>
> $$+11.9 - 6.02 - 13 + 10.93 + 4.31 - 3.37 = +4.75 \text{ dB}$$
>
> The output level is therefore
>
> $$0 + 4.75 \text{ dBm} = 4.75 \text{ dBm (an overall gain)}$$

Now go back to Figure 1.2 and verify for yourself that the decibel gains and losses given in Figure 1.3 are the equivalent power ratios used in Figure 1.2. Make a note that 4.75 dBm corresponds to a power of 2.98 mW, which is the output power of the first example if the input is set to 1 mW.

VOLTAGE GAIN   In general, two powers P_1 and P_2 will be compared by measuring either the voltage developed across a given impedance or the current flowing through it. If then the input and output impedances of a network are *equal*, the power ratio will be proportional to the square of the voltage or the current ratio. In such cases the impedances will cancel out and we can write

$$\frac{P_2}{P_1} = \left(\frac{V_2}{V_1}\right)^2 = \left(\frac{I_2}{I_1}\right)^2$$

Hence

$$\text{voltage gain} = 10 \log \left(\frac{V_2}{V_1}\right)^2 = 20 \log \frac{V_2}{V_1} \text{ dB}$$

$$\text{current gain} = 10 \log \left(\frac{I_2}{I_1}\right)^2 = 20 \log \frac{I_2}{I_1} \text{ dB}$$

It is very unusual for the input and output impedances of a network to be exactly equal, and it is strictly incorrect to describe voltage or current ratios in dB without regard to the necessary condition of such equality. It is not necessarily correct, for example, to say that if an amplifier has a voltage gain of 100, its power gain is 40 dB. This would be true only if the power gain was $(100)^2$ or 10 000.

The next example illustrates a typical case where the input and output impedances of a system are not equal.

> Example 6
> In a transistor amplifier, an input of 10 mV produces an output current of 1.5 mA in a 470 Ω output load resistor. If the input resistance of the amplifier is 600 Ω, find the voltage, current and power gains in dB.
>
> First
>
> $$\text{output voltage } V_o = 1.5 \times 10^{-3} \times 470 = 0.705 \text{ V}$$

$$\text{voltage gain } A_v = \frac{V_o}{V_i} = \frac{0.705}{10 \times 10^{-3}} = 70.5$$

Therefore in decibels,

$$A_v = 20 \log 70.5 = 37 \text{ dB}$$

Next,

$$\text{input current } I_i = \frac{V_i}{R_i} = \frac{10 \times 10^{-3}}{600} = 16.7 \text{ } \mu\text{A}$$

$$\text{current gain } A_i = \frac{I_o}{I_i} = \frac{1.5 \times 10^{-3}}{16.7 \times 10^{-6}} = 90$$

Therefore in decibels,

$$A_i = 20 \log 90 = 39 \text{ dB}$$

Finally,

$$\begin{aligned} \text{power gain} = A_v A_i &= 70.5 \times 90 = 6345 \\ &= 10 \log 6345 \text{ dB} \\ &= 38 \text{dB} \end{aligned}$$

Notice that the input and output resistances are roughly of the same order here. The corresponding dB ratios are not vastly different for all three forms.

PROBLEMS

1 Express as decibel gains or losses the voltage ratios 100, 40, 5 and 0.5. What power gains correspond to 20 dB, 7 dB, 6 dB and −30 dB?

2 What do you understand by the unit dBm? What special advantages does it have in the calculation of power ratios? What powers are represented by (a) 25 dBm (b) −13 dBm?

3 An amplifier has resistive input and output impedances each of 75 Ω. When an input signal of 500 mV is applied to the input terminals, a 10 V signal is developed across the output load. Calculate (a) the voltage gain ratio (b) the power gain ratio (c) the gain in decibels.
   If the output impedance is changed to 150 Ω resistive, the input impedance and the voltage gain ratio being unaltered, calculate (d) the power gain ratio (e) the gain in dB.

4 Show that a 3 dB voltage reduction is equivalent to a fall in the voltage to 0.707 of its original value.

5 A low-frequency amplifier has a power gain of 56 dB. The input impedance is 600 Ω resistive and the output load is 10 Ω resistive. What will be the current in the load when a 1 V RMS signal is applied at the input?

6 An amplifier has a response at 100 Hz which is 8.5 dB down on its response at 1 kHz. If the voltage gain ratio at 100 Hz is 15, what is the voltage gain at 1 kHz?

7 Define the decibel.

The input signal to an amplifier varies between 23.5 mW and 1.25 W. Express each of these powers in dB relative to 1 mW and calculate the fluctuation in the level of the signal in dB.

8 Voltmeters are available in which the scale is calibrated in volts and also in dB referred to 1 mW in 600 Ω resistive impedance. If the power in a line is measured by one of these meters and found to be 1 mW, what is the corresponding line voltage reading?

2 Principles of negative feedback

Aims: At the end of this unit section you should be able to:
Understand the basic principles of negative feedback.
Differentiate between series- and parallel-derived feedback methods.
Explain the effect of negative feedback on amplifier gain, bandwidth, noise, distortion and input and output resistance.
Apply feedback principles to practical amplifier circuits.

Many of the characteristics of amplifiers which we often treat as fixed and unvarying, such as gain and gain stability, bandwidth, and input and output resistances, are actually at the mercy of variations in circuit parameters and environmental conditions. Transistor gain tolerances, operating point shift, the ageing of components, temperature, climatic changes and supply voltage variations, to name only a few, all tend to and do affect an amplifier adversely.

The effects of most of these variations can be eliminated or at least considerably reduced by the introduction of controlled *feedback* which, as the name suggests, involves the addition of a portion of the amplifier output signal to the input signal. When the input signal is effectively reduced in magnitude by the addition of the feedback, the method is known as *degenerative* or *negative* feedback. If there is an increase in the total input due to the feedback, the method is termed *regenerative* or *positive* feedback. If an amplifier is badly designed, unintentional feedback can occur because of coupling between electric or magnetic fields in different parts of the circuit. Such feedback is invariably of the regenerative kind and the amplifier is quite unsuitable for its intended function in life. Controlled feedback can only be usefully applied to an amplifier which has been soundly designed in the first place.

The use of negative feedback is widespread, but it must be used with caution. Negative feedback significantly modifies the characteristics of an amplifier to which it is applied. It enables a predictable performance to be achieved, free from the effects of the variations in the circuit components and devices from which it is constructed. It also ensures repeatability; every amplifier in a manufactured batch has identical characteristics and lies within tight specification limits.

We shall be concerned in this section with amplifier systems in which the output signal is fed back to the input terminals in antiphase with the input signal. This is negative feedback (NFB). Cases of positive feedback in which the signal is fed back in phase with the input will be covered in Section 4.

The manner in which the output signal is sampled and the manner in which it is introduced into the input circuit lead to four basic classifications of NFB amplifiers. Two of these are shown in *Figure 2.1(a)* and (*b*). These illustrate those cases

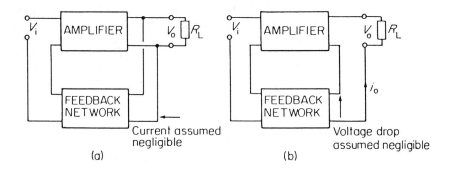

Figure 2.1

where the voltage fed back is applied in *series* with the input. In diagram (*a*) the voltage fed back from the output is proportional to the output voltage V_o and is derived from a network in parallel with the load resistor R_L; this is known as *voltage–voltage* feedback. In diagram (*b*) the voltage fed back and applied in series with the input is proportional to the current flowing in the load resistor, i_o. This is known as *voltage–current* feedback, although it is a voltage and not a current which is actually fed back.

We consider first the general principles involved when the feedback is applied in series with the input signal, irrespective of the form of connection at the output end.

VOLTAGE FEEDBACK PRINCIPLES

In voltage feedback, the signal fraction fed back is proportional to the voltage of the signal at the output of the amplifier, or at some point within the amplifier at which the feedback signal is derived. *Figure 2.2(a)* shows an amplifier with voltage gain A_v supplying a load resistor R_L. As we know,

$$\text{voltage gain } A_v = \frac{V_o}{V_i}$$

or

$$V_o = A_v V_i$$

Figure 2.2(b) illustrates the same amplifier with voltage feedback. The basic amplifier gain is still A_v and the output

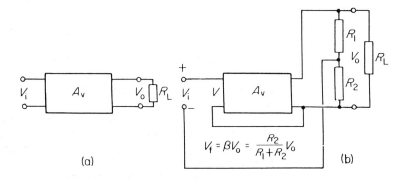

Figure 2.2

voltage V_o is developed as before across R_L. In parallel with R_L is a feedback network consisting of two series-connected resistors R_1 and R_2. We assume that the total resistance of R_1 plus R_2 is so large compared with R_L that the current through them is negligible. That part of the output voltage developed across R_2 is fed back and added in series with the input voltage. Let this fraction be called β, so that

$$\beta = \frac{R_2}{R_1 + R_2}$$

Then the voltage fed back $V_f = \beta V_o$, and the total input voltage now available at the *terminals* of the amplifier is

$$V = V_i + \beta V_o$$

But now

$$A_v = \frac{V_o}{V} \quad \text{or} \quad V = \frac{V_o}{A_v}$$

Therefore

$$\frac{V_o}{A_v} = V_i + \beta V_o$$

and so

$$V_o = A_v V_i + A_v \beta V_o$$

Hence the gain with the feedback applied, measured between the *new* input terminals and the output, is

$$\frac{V_o}{V_i} = A_v' = \frac{A_v}{1 - \beta A_v} \qquad (2.1)$$

This is the *general feedback equation*.

The quantity βA_v in this equation is called the *loop gain* since it is the total gain measured round the feedback loop from V to V_f. That is, it is the product of the individual gains of the amplifier A_v and the feedback network β. In the absence of feedback, $\beta = 0$ and the amplifier exhibits a gain equal to A_v. This is the *open-loop* gain.

There are three possible forms that the loop gain product may take which are of interest:

1 If βA_v lies between 0 and 1 (but not equal to 1), the denominator of equation (2.1) will be less than unity and the gain with feedback Av' will be greater than A_v. This is a case of *positive* feedback because the gain of the system is increased.
2 If BA_v is negative, the denominator will be greater than unity since $1 - (-\beta A_v) = 1 + \beta A_v$, and A_v' will be less than A_v. This is a case of *negative* feedback because the gain of the system has been reduced.
3 If $A_v = 1$ the denominator becomes zero and the resultant gain is theoretically infinite. This implies that the circuit has an output independent of an external input voltage; hence the circuit will be unstable and *oscillate*. We shall consider this special condition in Section 4. Strictly, equation (2.1) does not

apply in this case, since it was derived on the assumption that the amplifier system of Figure 2.2(b) was stable.

The general feedback equation is very simple to use and is easily remembered.

Example 1

In a certain amplifier the open-loop gain $A_v = 200$. Find the overall gain with feedback when (a) $\beta = 0.004$ (b) $\beta = -0.02$.

(a) $A_v' = \dfrac{A_v}{1 - \beta A_v} = \dfrac{200}{1 - (0.004 \times 200)} = \dfrac{200}{0.2} = 1000$

(b) $A_v' = \dfrac{200}{1 - (-0.02 \times 200)} = \dfrac{200}{5} = 40$

Clearly the introduction of negative feedback, as part (b) of this example illustrates, has led to a drastic reduction in gain. This will always be the case if βA_v is negative. This might seem at first to be a considerable handicap, but the advantages of feedback are so numerous that it pays to design the original amplifier with sufficient gain that, after feedback has been added, the overall gain will not fall below the desired figure.

PHASE-SHIFT REQUIREMENTS

Let us take a more general view of an amplifier and a feedback network; this is illustrated in *Figure 2.3*. For βA_v to be negative, either A_v or β must be negative. Now either the amplifier or the feedback network, or both, may introduce a phase shift between their input and output terminals. These phase shifts have been respectively indicated by the angles θ and ϕ in Figure 2.3.

In our previous diagram of Figure 2.2(b) the feedback network consisted simply of a resistance divider; hence there was no phase shift there, and angle ϕ was zero. For βA_v to be negative, therefore, the amplifier gain must be $-A_v$; hence the amplifier phase shift θ must be 180°. The total phase shift around the loop is then 180° and so the voltage fed back to the input (V_f) is antiphase to the existing input. The signal polarity signs marked on Figure 2.3 show how this occurs.

If the amplifier consisted, say, of two common-emitter stages, its total phase shift would be 360° which is equivalent to a zero phase shift; hence $\theta = 0°$. In order to provide negative feedback for this amplifier, the phase shift ϕ in the feedback network would have to be 180°.

To simplify matters we will assume throughout this section that βA_v is negative, and not worry ourselves whether it is the amplifier or the feedback network which is actually contributing the sign. For negative feedback, therefore, we can take the feedback equation to be

$$A_v' = \frac{A_v}{1 + \beta A_v}$$

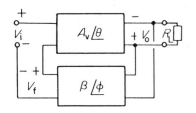

Figure 2.3

GAIN STABILITY As we have seen, amplifier gain can be affected by changes in the supply voltage, the ageing of components (particularly valves), and the replacement of active devices (particularly transistors) with others which do not have the same characteristics as the originals. Negative feedback can do much to counteract the effects of all these variations. An example will be a useful introduction to the case of gain stability.

Example 2
An amplifier with a voltage gain of 20 000 is used in a feedback circuit where $\beta = 0.02$. Calculate the overall gain. If the gain dropped by one-half of its inherent value because of supply voltage variation, what would the overall circuit gain now become?

With the full gain available

$$A_v' = \frac{20\ 000}{1 + (0.02 \times 20\ 000)} = 49.9$$

When the gain falls to 10 000, the new gain with feedback is

$$A_v' = \frac{10\ 000}{1 + (0.02 \times 10\ 000)} = 49.75$$

This example illustrates a very important point: although the inherent gain of the amplifier without feedback fell from 20 000 to 10 000, a change of 50 per cent, the overall gain of the system with feedback dropped only from 49.9 to 49.75, a fall of less than 0.5 per cent. The circuit with feedback has therefore a more stable gain figure than the amplifier alone. If we examine the general feedback equation

$$A_v' = \frac{A_v}{1 + \beta A_v}$$

then, when A_v is very much greater than 1,

$$A_v' \simeq \frac{A_v}{\beta A_v} \simeq \frac{1}{\beta}$$

This makes A_v' independent of A_v, since only β is involved in this expression. Hence the gain of a feedback amplifier is very stable.

Since β is fractional, the way to make βA_v large is to make A_v large by starting off with a very high-gain amplifier. This is not always an easy option, as high gain tends to introduce instability because of undesirable positive feedback occurring between the circuit wires and component fields. Negative feedback will do nothing to put right a poor initial design.

1 An amplifier has a gain of -1500 and is used with a feedback network where $\beta = 0.1$. Calculate the overall voltage gain.
2 An amplifier has a gain of 10^3 without feedback. Calculate the gain when 0.9 per cent of negative feedback is applied.
3 You require an amplifier with a final gain of 75. What should be the open-loop gain of the basic amplifier if the feedback fraction is to be -0.01?
4 An amplifier having an open-loop gain of 500 has overall NFB applied which reduces the gain to 100. What is the feedback fraction?
5 What would the gain of an amplifier become if the whole of the output voltage was fed back in opposition to the input? Is such a design possible?

ADVANTAGES OF NEGATIVE FEEDBACK

We will now work through a number of advantages that result from the application of NFB to an amplifier.

Reduction of distortion

Any distortion which may be present in the *original* signal input to an amplifier is indistinguishable from the signal and is therefore unaffected by NFB. Distortion may also arise *within* the amplifier, for example a harmonic introduced in the output stage. If a portion of this is fed back in opposition to the generated harmonic, the amount of distortion will be reduced.

Let an amplifier without feedback be considered as having harmonic distortion introduced by means of a hypothetical voltage generator as shown in *Figure 2.4(a)*. The distortion fraction is then D/V_o, where D is the distortion voltage magnitude. Now suppose a feedback network is added to the amplifier as shown in *Figure 2.4(b)*. Since the distortion usually

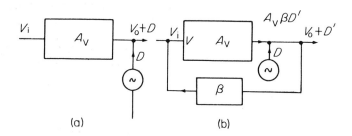

Figure 2.4

depends upon the magnitude of the input voltage V, suppose V_i to be so adjusted that the input voltage at the amplifier terminals is the same as before. Then the harmonic voltage fed back is $\beta D'$, say, and this appears at the output as $A_v\beta D'$. Hence

$$D' = D + A_v\beta D'$$

Therefore

$$\frac{D'}{D} = \frac{1}{1 - \beta A_v}$$

$$= \frac{1}{1 + \beta A_v} \quad \text{for } \beta A_v \text{ negative}$$

So the distortion fraction is improved by the factor $1/(1 + \beta A_v)$.

This argument may seem academic at first, for the distortion is reduced by the same factor as is the gain of the system. However, harmonic distortion usually results from large signals traversing the extremes of the dynamic output characteristics, and such excursions most often occur in the power output stages of amplifiers. Using NFB over such stages to reduce the distortion and to recover the lost gain in the small-signal stages earlier in the amplifier is a common practice. You will probably see a 'catch' at this point. A logical question to ask is; what happens to the harmonic distortion when we increase the gain to what it was originally? It might be best to answer this by way of a worked example.

Example 3
An amplifier has an open-loop gain of 100 and is used with a feedback network where $\beta = -0.1$. If the fundamental output voltage is 10 V with 20 per cent second-harmonic distortion, calculate (a) the input required to maintain the output at 10 V (b) the value of second-harmonic voltage when feedback is applied (c) the total overall distortion when the gain is restored to its original value.

(a) With feedback,

$$A_v = \frac{100}{1 + (0.1 \times 100)} = \frac{100}{11} = 9.1$$

Hence to provide 10 V output with this gain, the required input will be $10/9.1 = 1.1$ V.

(b) Since there is 20 per cent distortion, the distortion voltage at the output without feedback is 20 per cent of 10 V, that is 2 V. Then with feedback the distortion voltage is reduced to

$$D' = \frac{D}{1 + \beta A_v} = \frac{2}{11} = 0.182 \text{ V}$$

Notice that the gain and the distortion have both been reduced by the same factor, i.e. 1/11.

(c) We now want to restore the gain to its original value of 100. Since the gain with feedback is 9.1, we shall need *three* such stages to obtain a gain of at *least* this figure.

In the original amplifier the percentage harmonic was 20 per cent and with feedback this drops to $0.182/10 \times 100$ per cent. The first stage will now introduce 1.82 per cent harmonic distortion and this will be fed as input to the second stage. This will add a further 1.82 per cent harmonic of its own, and so the total harmonic will be 3.64 per cent at the output of the second stage. (Make certain you understand why we *add* the distortion percentages.) In the same way the third stage will contribute a further 1.82 per cent and the total distortion at the output will be 5.46 per cent.

This is a considerable improvement on the original 20 per cent. Actually, with three stages we have more gain than we need; the signal level in the first two stages could be kept low and the overall harmonic content at a gain of 100 would be much less than the figure of 5.46 per cent calculated.

Noise Consider the diagram of *Figure 2.5*, which is the same as that shown in Figure 2.1(*b*) but with an additional input signal N which can be assumed to be a noise voltage introduced into the first stage of the amplifier. Then the input signal becomes

$$V = (V_i + N) + \beta V_o$$

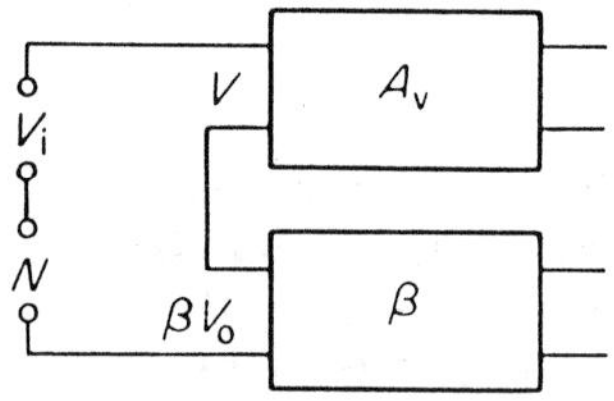

This equation indicates that the noise signal will be dealt with in the same manner as V_i; that is, N will be reduced by the same amount as V_i. Hence

$$N' = \frac{N}{1 + \beta A_v}$$

Noise may be reduced by the application of NFB but, because of the requirement for very high forward gain and so an increased number of stages of amplification, it is possible for the overall noise level to be increased. Once again, NFB is not a panacea for a poor original design.

Figure 2.5

6 Can NFB have any effect on signal-to-noise ratio?

Increase in bandwidth For large amounts of NFB, $A_v' \simeq 1/\beta$, hence if β can be made independent of frequency, the overall gain will be independent of frequency. It is not possible to obtain an infinite bandwidth by this method (or any other) but NFB can nevertheless extend the bandwidth of an amplifier by a considerable amount.

As we have already noted, bandwidth is defined as that range of frequencies over which the gain does not fall below $1/\sqrt{2}$ of its maximum (usually its mid-band) value. *Figure 2.6* shows the open-loop response of an amplifier with a bandwidth $f_2 - f_1$ Hz and mid-frequency gain A_o. When NFB is applied this gain is reduced to $A_o' = A_o/(1 + \beta A_o)$. A study of the diagram makes it clear that the bandwidth has now increased to $f_2' - f_1'$. It can be proved that

$$f_2' = f_2(1 + \beta A_o)$$

and

$$f_1' = \frac{f_1'}{1 + \beta A_o}$$

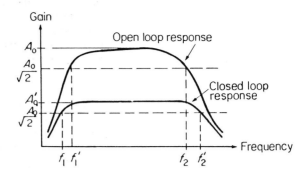

Figure 2.6

As f_1' is often reduced to a very low frequency, to a good approximation the overall bandwidth with large amounts of NFB is f_2'.

A *figure of merit* for a wideband amplifier is its gain–bandwidth product. Negative feedback reduces the gain but increases the bandwidth; overall the figure of merit may be relatively unaffected.

> 7 The response curve of an amplifier is given in *Figure 2.7*. Calculate the resultant values of gain at 100 Hz, 1 kHz, 10 kHz and 100 kHz when 1 per cent of the output voltage is fed back in antiphase to the input. Sketch (on the same axes) the resultant response curve.

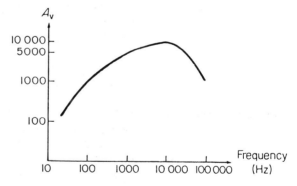

Figure 2.7

CURRENT FEEDBACK PRINCIPLES

Before going on to discuss the effect of NFB on input and output resistances, we will consider the cases of feedback amplifiers where the signal fed back is applied in *parallel* with the input terminals. In these cases we are concerned with the addition of currents, not voltages, and so these forms of feedback are given the names *current–current* feedback or *current–voltage* feedback, depending upon the manner in which the signal is derived at the output end.

Figure 2.8(a) and (*b*) show respectively current–current feedback and current–voltage feedback. By simple consideration, parallel-connected feedback will *not* affect the

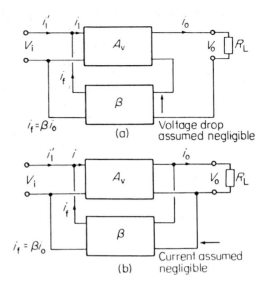

Figure 2.8

voltage gain of the amplifier, since for a given voltage output the *same* input voltage is required whether or not the feedback circuit is connected. However, the current gain is affected, for from Figure 2.8(a)

$$i_f = \beta i_o$$

$$i_1 = i' + i_f$$

$$i_1' = \frac{i_1}{1 + \beta A_i}$$

where A_i is the current gain ratio i_o/i_1. Then

$$A_i' = \frac{A_i}{1 + \beta A_i}$$

taking βA_i as being negative. This is analogous to the series voltage feedback cases with current gain A_i replacing voltage gain, and β being the current ratio i_f/i_o.

This relationship is not of such interest to us. Most amplifiers are designed as voltage amplifiers, and for those using FET or valve input stages the current amplification has little meaning.

We now investigate the effect of NFB on the input and output resistance of amplifiers.

INPUT RESISTANCE

The input resistance of an amplifier is the resistance as measured at the input terminals. In the following paragraphs we shall consider only the case of amplifiers which, before feedback is applied, have purely resistive input and output impedances. For most amplifiers, this is a reasonably true state of affairs.

There are two cases to consider from the point of view of the effect of NFB on input resistance: (a) feedback is in series, (b) feedback is in parallel with the input.

Series feedback

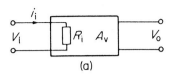

(a)

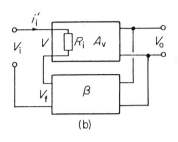

(b)

Figure 2.9

Figure 2.9(a) shows an amplifier without feedback, of open-loop gain A_v and input resistance R_i. This resistance can be measured as being simply the ratio of V_i to i_i. In *Figure 2.9(b)* feedback has been applied in series with the input. Now since V_f is opposing V_i, it is evident that the effective voltage V at the input terminals will now be less than V_i and hence, as seen from these terminals, the input current will be *reduced* from i_i to i_i'. This means that the input resistance as seen from the *source* terminals has apparently *increased*. Let us check this out with a little algebra.

The input resistance with feedback will be

$$R_i = \frac{V_i}{i_i}$$

and

$$V_i + V_f = i_i'R_i \tag{2.2}$$

Now

$$V_o = A_v'V_i = V_i\frac{A_v}{1 + \beta A_v}$$

Therefore

$$V_f = \beta V_o$$

$$= \frac{\beta A_v V_i}{1 + \beta A_v}$$

Substituting this value of V_f into (2.2):

$$V_i + \frac{\beta A_v V_i}{1 + \beta A_v} = i_i R_i$$

$$V_i\left(1 + \frac{\beta A_v}{1 + \beta A_v}\right) i_i'R_i$$

But with feedback,

$$R_i' = \frac{V_i}{i_i'} = R_i(1 + \beta A_v)$$

Hence, as we deduced, the input resistance is increased with series applied NFB by a factor $(1 + \beta A_v)$.

Parallel feedback

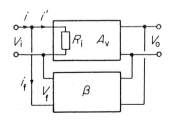

Figure 2.10

We now consider the situation where the feedback is applied in parallel with the input as shown in *Figure 2.10*. In the parallel circuit, as mentioned, we are concerned with current addition, not voltage. The input current i is the sum of the amplifier input current i' and the feedback current i_f. So the input resistance R_i' with feedback applied is

$$R_i' = \frac{V_i}{i' + i_f} \quad \text{where} \quad i_f = \frac{V_f}{R_i} \quad \text{and} \quad i' = \frac{V_i}{R_i}$$

$$\therefore R_i' = \frac{V_i}{(V_i/R_i) + (V_f/R_i)} = \frac{V_iR_i}{V_i + \beta A_v V_i}$$

$$R_i' = \frac{R_i}{1 + \beta A_v}$$

So with parallel-connected feedback the input resistance is *reduced* by a factor $1/(1 + \beta A_v)$

Example 4

An amplifier has an open-loop gain of 5×10^5 and an input resistance of 100 kΩ. When NFB is applied in series with the input, the gain is reduced to 10^3. Find (a) the value of the feedback fraction β (b) the new input resistance.

(a) $A_v' = \dfrac{A_v}{1 + \beta A_v}$

$\qquad A_v = A_v' + \beta A_v A_v'$

$\qquad \beta = \dfrac{A_v - A_v'}{A_v A_v'} = \dfrac{(5 \times 10^5) - 10^3}{5 \times 10^8} = 0.998 \times 10^{-3}$

(b) For series connection,

$\qquad R_i' = R_i(1 + \beta A_v)$

$\qquad = 100 \times 10^3 \, [1 + (0.998 \times 10^3 \times 5 \times 10^5)] \quad \Omega$

$\qquad = 100 \times 10^3 \, [1 + (4.99 \times 10^2)] \quad \Omega$

$\qquad = 100 \times 10^3 \times 500 \quad \Omega$

$\qquad = 50 \text{ M}\Omega$

OUTPUT RESISTANCE

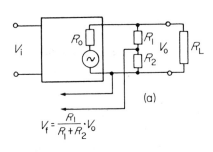

$$V_f = \frac{R_1}{R_1 + R_2} \cdot V_o$$

(a)

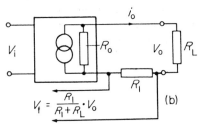

$$V_f = \frac{R_1}{R_1 + R_L} \cdot V_o$$

(b)

Figure 2.11

The output resistance of an amplifier is the resistance measured at the output terminals when the input is zero. There are again two cases to consider when feedback is taken from the output terminals.

Figure 2.11(a) illustrates the feedback network connected in parallel with the output, the method shown in Figure 2.2(b). This is the case of voltage feedback, because V_f is directly proportional to V_o, and β (the feedback fraction) is given as $R_2/(R_1 + R_2)$. It is assumed that the total resistance of R_1 and R_2 in series is very much greater than the load R_L, so that the introduction of the feedback network does not affect the loading.

Figure 2.11(b) illustrates current feedback; here the load current is passed through a resistor R_1 in series with load R_L. The feedback quantity is now proportional to the output current i_o and the feedback fraction β is $R_1/(R_1 + R_L)$.

R_1 is very small compared with R_L, so that the load current is unaffected by its introduction. β then approximates to R_1/R_L.

The change in output impedance with feedback may be deduced qualitatively by considering the diagrams in turn. Suppose in (a) V_o increases for some reason but V_i remains constant. V_f will increase, but as the feedback is negative the effective input voltage will be reduced and this in turn will tend to reduce the output voltage. The net effect is to keep V_o

constant; hence the output circuit acts as a device with a *small* internal resistance, that is as a constant-voltage source. The diagram illustrates this form of output equivalent circuit.

Conversely, in diagram (b), if i_o increases, V_f increases and the effective input quantity (current or voltage) will decrease. The net effect is to keep i_o constant and so the output circuit behaves this time as a constant-current generator, that is a device with a *large* output resistance. This is shown on the diagram.

Hence we deduce that voltage-derived (parallel) feedback reduces the output resistance, and current-derived (series) feedback increases the output resistance. You must not assume that these changes in output resistance come about because the parallel feedback resistors are shunting the load, or that the series feedback resistor adds to the load. We have already stated that the parallel resistor draws negligible current and the series resistor drops negligible voltage.

It can be proved that for the parallel connection

$$R_o' = \frac{R_o}{1 + \beta A_v}$$

and for the series connection

$$R_o' = R_o(1 + \beta A_v)$$

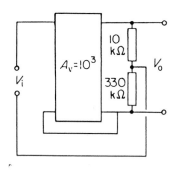

Figure 2.12

Example 5

In the feedback circuit shown in *Figure 2.12*, the amplifier gain is 10^3 whent he 330 Ω resistor is short-circuited. What is the gain when the short-circuit is removed? If the resistors used have a tolerance of ±5 per cent, calculate the maximum and minimum values that the gain might have with feedback.

When the 330 Ω resistor is short-circuited, there is no feedback and so the gain $A_v = 1000$.

When the 330 Ω resistor is in circuit, feedback is applied and the feedback fraction $\beta = 0.33/(10 + 0.33) = 0.032$. Therefore

$$A_v' = \frac{A_v}{1 + \beta A_v} = \frac{1000}{1 + (0.032 \times 1000)} = 30.3$$

If the resistors are ±5 per cent tolerance, we must find the extreme values that β can take. Assuming that the 330 Ω resistor is at the top end of its tolerance range and the 10 kΩ resistor is at the bottom end of its range, their respective values are 346.5 Ω and 9.5 kΩ. Then

$$\beta = \frac{0.3465}{9.5 + 0.3465} = 0.0352$$

$$A_v' = \frac{1000}{1 + 35.2} = 27.6$$

when the 330 Ω resistor is at the bottom end of its tolerance range and the 10 kΩ resistor is at the top end, their respective values are 313.5 Ω and 10.5 kΩ. Then

$$\beta = \frac{0.3135}{10.5 + 0.3135} = 0.029$$

$$A_v' = \frac{1000}{1 + 29} = 33.34$$

The gain could therefore lie between the limits 27.6 and 33.34.

Now try these two problems on your own.

---

8 If the gain of an amplifier without feedback is 55 dB, what will be the resultant gain in dB when negative feedback is applied from the output to the input and $\beta = 0.1$?

9 An amplifier has an open-loop gain of 300, which is found to fall by 20 per cent due to changes in the supply voltage. If the gain is to be stabilized so that it falls by only 1 per cent, calculate the required value of feedback fraction β. If the original bandwidth of this amplifier was 500 Hz to 50 kHz, calculate the bandwidth when feedback is applied.

INSTABILITY IN NEGATIVE FEEDBACK AMPLIFIERS

On several occasions throughout this section, mention has been made of the necessity of careful design when negative feedback is being applied to an amplifier system. It is often put about that negative feedback is used to prevent oscillation in the amplifier, but this is quite wrong. Indeed, negative feedback carelessly applied will often cause a perfectly stable amplifier to become unstable.

One of the major problems associated with negative feedback is that of phase shift around the loop. At the so-called mid-frequencies of the amplifier passband, approximating to the normal 3 dB bandwidth range, there is a 180° phase shift in each stage of a voltage amplifier, and the feedback voltage is arranged to be in phase opposition to the input signal. At frequencies beyond the passband, the reactive components within the amplifier introduce additional phase shifts, so that the total phase shift moves from 180° down towards 90° or up towards 270°. It is then possible for the feedback to become positive at certain low or high frequencies, the feedback signal adding to the input instead of opposing it. The result is that the circuit will become unstable and oscillate, so invalidating its function as an amplifier (or at least as an amplifier whose parameters are under our control). This trouble can only be eliminated by ensuring that the loop gain βA_v is less than unity at the frequencies in question.

THE EMITTER FOLLOWER

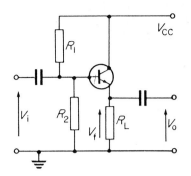

Figure 2.13

The effective voltage across
the base-emitter
junction is V_i-V_f

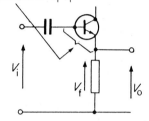

Figure 2.14

The common-collector circuit of *Figure 2.13* – or, as it is more usually known, the *emitter follower* amplifier – is an example of voltage–voltage feedback. The load R_L is in the emitter circuit, resistors R_1 and R_2 are the usual bias components, and the collector is returned directly to the V_{CC} line where, from the signal point of view, it is effectively at earth potential. Assume that we apply a sinusoidal signal at the input terminals.

As V_i rises towards a maximum, emitter current will increase and so the emitter voltage will rise relative to earth. The converse will happen when the input goes negative. Hence the emitter voltage 'follows' the input voltage, and V_i and V_o are *in phase* relative to the earth line. However, in the base circuit input relative to the *emitter*, V_i and V_f are in series. Hence the effective base–emitter potential is the difference between V_i and V_f, and these voltages are therefore in *phase opposition* (*Figure 2.14*). Notice that in this circuit V_o is in phase with V_i and hence $A_v = V_o/V_i$ is positive; but V_f is phase opposed to V_i and hence to V_o, and therefore $\beta = V_f/V_o$ is negative. But $V_f = V_o$; hence $\beta = -1$. This means that *all* the output voltage is fed back in series with the input, and 100 per cent negative feedback is applied.

Substituting $\beta = -1$ in the feedback equation (2.1) we get

$$A_v' = \frac{A_v}{1 - \beta A_v} = \frac{A_v}{1 + A_v}$$

This must be less than unity; hence the gain of the emitter follower is always slightly less than 1 and there is no voltage phase reversal.

The usefulness of the emitter follower stems from the effect that 100 per cent NFB has on the input and output resistances. As the circuit is voltage–voltage feedback, with the feedback derived in parallel with the output load and fed back in series with the input circuit, the respective resistances can be expected to be high at the input and low at the output terminals. Hence the amplifier is ideal as an 'electronic transformer' for matching a stage of high-resistance output into one of low-resistance input without the voltage reduction or frequency restriction inherent in an ordinary step-down transformer. The low output resistance is also useful when pulse signals are fed to a capacitive device such as a cathode-ray tube, distortion of the pulses being greatly reduced.

It can be proved that to a good approximation

$$R_i' \simeq h_{fe}R_L$$

(but account has to be taken of the shunting effect of the bias resistors), and

$$R_o' \simeq \frac{R_S}{h_{fe}}$$

where R_S is the resistance of the signal source feeding the amplifier. This value of R_o' has R_L effectively in parallel with it.

BOOTSTRAPPING

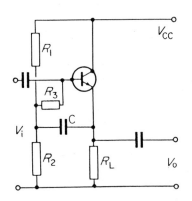

Figure 2.15

It is the shunting effect of the bias resistor(s) in the emitter follower that degrades the very high input resistance that this amplifier otherwise provides. By arranging the circuit in the manner shown in *Figure 2.15* the paralleling effect of bias resistors R_1 and R_2 upon the input is practically eliminated. This is known as *bootstrapping*.

Capacitor C acts as a DC block between the emitter and base circuits but is selected to have a negligible reactance at the signal frequency. Hence feedback occurs between the output signal at the emitter terminal and the input signal at the base through this capacitor. Because $A_v \simeq 1$, the signal amplitude across R_L is almost of the same amplitude as the incoming signal; in addition, it is in phase with the incoming signal. Hence the potentials at any instant at either end of resistor R_3 are approximately equal. This is equivalent to R_3 having a very high resistance to the signal frequency but *not* to the DC bias current feeding the base. Hence the shunting effect of R_1 and R_2 is isolated from the input terminals.

10 Would it help if the value of R_3 in the bootstrap circuit (Figure 2.15) was made very large, say several megohms?

11 Define the terms (a) open-loop gain (b) feedback fraction (c) gain stability.

12 Say whether each of the following statements is true or false:
Series-connected negative feedback
(a) increases the input resistance
(b) reduces the voltage gain
(c) increases the current gain.
Parallel-connected negative feedback
(d) increases the output resistance
(e) leaves the voltage gain unaffected
(f) reduces the current gain.

13 The voltage gain of an amplifier is A. If a fraction β of the output voltage is fed back in antiphase with the input, derive an expression for the new overall gain of the amplifier.

14 In a certain amplifier $A_v = 50$. Find the value of A_v' if (a) $\beta = 0.005$ (b) $\beta = 0.02$ (c) $\beta = -0.02$. Comment on each of your solutions.

15 A transistor provides a gain of 50. If feedback is applied so that the gain increases to 62.5, what sort of feedback is it, and what fraction of the output is fed back?

16 An amplifier has 0.5 per cent of its output voltage fed back in antiphase to the input. If the gain of the amplifier is then 160, what was its gain before feedback was applied?

17 An amplifier has an open-loop gain of 1000, which is likely to vary by 10 per cent because of supply fluctuations. If feedback with $\beta = -0.01$ is applied to the amplifier, calculate the overall variation in gain.

18 An amplifier with a voltage gain of 2×10^3 is used in a feedback arrangement where $\beta = -0.02$. Calculate the

overall gain. If the gain dropped to one-half of its inherent value, what would the overall circuit gain become?

19 An amplifier has a voltage gain of 57 dB. When NFB is applied the gain falls to 27 dB. Calculate β.

20 An amplifier has a voltage gain of 60 dB without feedback and 30 dB when feedback is applied. If the gain without feedback changes to 55 dB, calculate the new gain with feedback.

21 Negative series feedback is applied to an amplifier to reduce its open-loop gain by 50 per cent. By what factor will the input and output resistances change?

22 An amplifier has a voltage gain of 60 dB. Find the dB change in the gain if 1/30 of the output signal is fed back to the input in phase opposition. Ignoring any phase change in either the amplifier or the feedback network, calculate the reduction in harmonic distortion at the amplifier output.

23 An amplifier circuit is to be designed to have an overall gain of 20 ± 0.1 per cent. The basic amplifier from which the final circuit is to be derived has a gain $A_v \pm 10$ per cent. Calculate the required value of β and the gain A_v.

24 A NFB amplifier is shown in *Figure 2.16*. When the 1 kΩ resistor is short-circuited the voltage ratio V_o/V_i is 2000. What will this ratio be when the short-circuit is removed?

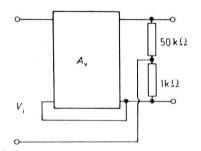

Figure 2.16

3 Unwanted outputs: noise

Aids: At the end of this unit section you should be able to:
Define noise as any unwanted signal.
List the sources of internal and external noise and explain how they occur.
Define signal-to-noise ratio and noise factor in an amplifier or receiver, and perform simple calculations relating to these definitions.
State the precautions taken to minimize the effects of external noise.
Explain a method of measuring noise factor.

Disconnect the aerial from your radio receiver (or tune it to a point between stations), turn up the volume control and put your ear to the loudspeaker. In theory you should hear nothing but you will in fact hear a background hiss which, if you listen carefully enough, will be heard to vary in intensity in a completely random fashion. You will hear the same sort of thing from any amplifier system which feeds into a loudspeaker, even though the input to the amplifier is short-circuited to earth.

This background noise is present even when the amplifier is working normally, though it tends to be concealed behind the speech or music output. Unlike distortion, the noisy background is present whether the signal itself is there or not. Other noises are also often present. There may be a background hum breaking through from the AC supply to the amplifier. Crackling noises may be present, particularly when thunderstorms are in the vicinity of the receiver, or switches are operated in other rooms of the house. The list of such unwanted outputs is very long and all of them come under the heading of 'noise'. Noise may be visual as well as aural. Noise appearing on a television signal, for example, shows itself on the screen as flickering spots or bands of varying light intensity.

We define *noise* as any spurious or unwanted electrical signal set up in or introduced into an electronic system. Noise can be divided into two general classes:

1 Internal noise generated within the electronic system as the result of the random movement of charge carriers in resistors, wires and active devices such as transistors and valves.
2 External noise caused by atmospheric disturbances, diathermy apparatus, aircraft reflections and any spark-producing systems such as motor commutators or car ignition, to name just a few.

We will deal with these two classes in turn.

INTERNAL NOISE   Although noise will be of concern at the output terminals of an amplifier or receiver, whether this output feeds to a loudspeaker, cathode-ray tube or any device on which the information may be recorded or displayed, it is in the early or small-signal stages that internally generated noise is of importance.

You will find that if you turn down the volume control on an amplifier system (in the absence of any signal), the background noise will become insignificant; so the large-signal stages which generally follow the position of the volume control are not contributing much of the noise you hear when the volume is increased. The bulk of the noise is appearing *before* the volume control, that is, in the small-signal stages. So the input stage of any amplifier or receiver is, in general, the vital area for investigation into noise generation, for the noise set up at this point determines the useful operational sensitivity of the unit as a whole. If the input signal magnitude is less than the noise level, the output will be masked by the noise and become unintelligible.

Noise can be generated in any stage of an amplifier and from a variety of different processes. Many of these can be eliminated or minimized by careful attention to design and as such are not of primary interest to us in this section. For example, the proximity of an inductor, perhaps carrying 50 Hz AC to the input terminals of an amplifier, may introduce an unwanted hum signal into the amplifier. This is a case of stray inductive coupling, which can be completely eliminated by magnetically screening the offending inductor and/or shifting its position and orientation relative to the sensitive input terminals. Stray capacitive couplings can likewise lead to the same trouble, and can equally be eliminated by suitable screening and component disposition.

Other sources of internal noise are not, however, disposed of quite so easily. Internal noise is generated in resistors, transistors, valves and in fact in every piece of wire used in the construction of an amplifier, and is a function of the physical properties of the materials used and the environment in which they operate. As a result such noise is substantially beyond the control of the designer to do anything about, at least as far as layout is concerned.

Noise voltages of this sort are completely random in the sense that their future behaviour is unpredictable – in contrast to a sine wave, for example, whose past history is accurately known and whose future variations can be precisely stated. Noise voltages are not restricted to any particular frequency or phase or to any particular magnitude.

Thermal noise Thermal noise, which is also known as Johnson noise, is the noise associated with the random movement of electrons in any electrical conductor. These movements, caused by thermal agitation, are present in every conductor even in the absence of an association with an electric circuit.

A piece of wire resting on a bench is generating thermal noise voltages. As the temperature of a conductor is increased from absolute zero, 0 K (= −273°C), the atoms of the conductor begin to vibrate and more and more electrons are gradually released from the outermost orbits of the atoms to become free charge carriers. These free electrons wander about within the atomic structure of the material, first in one direction and then another, but at any given instant there will be more of them moving in one particular direction than another.

Since the movement of electrons constitutes an electric current, an EMF will be instantaneously induced between the ends of the conductor in the manner shown in *Figure 3.1(a)* and *(b)*. The time duration of each of these possible conditions is extremely short and so the polarity of the generated EMF is fluctuating rapidly, as is its magnitude, and in a completely random way. This is thermal agitation noise.

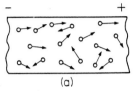 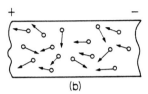

(a)	(b)
At this instant the direction of electron movement is predom- inantly left to right. The emf. is set up with the polarity shown	At this instant the direction of electron movement is predominantly right to left. The emf. is set up with the polarity shown

Figure 3.1

Noise voltage is measured in terms of RMS and is given by the relationship

$$V_n = \sqrt{(4kTRB)} \text{ volts} \qquad (3.1)$$

where

k is a constant $= 1.38 \times 10^{-23}$ joules/kelvin
T is the temperature of the conductor in kelvins
(K = °C + 273°)
R is the resistance of the conductor in ohms
B is the bandwidth of the circuit in which the conductor is situated.

This last factor requires some additional explanation. Where noise is concerned, we must be interested in the *total* bandwidth of the system, so the whole of the area enclosed by the response curve must be considered. *Figure 3.2* shows a typical response curve where, in the usual way, the power gain is plotted against frequency. The equivalent noise bandwidth is obtained by

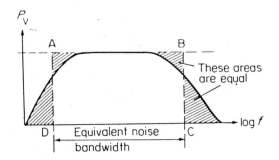

Figure 3.2

constructing a rectangle ABCD which has an area equal to the area under the actual response curve. This can be done by estimating equal areas as shown in the diagram. The resulting noise bandwidth is *not* the same as the bandwidth obtained at the 3 dB points, but it is often sufficiently accurate to take the latter measurement as being the same as the noise bandwidth. Noise voltage depends upon the bandwidth of the circuit over which the noise is measured (or of the circuit at whose output the noise is present), whichever is the smaller.

Thermal noise contains equal magnitudes of *all* frequencies and is known as *white noise*. It clearly follows that we should expect a wideband amplifier to generate more noise than a narrowband amplifier.

Example 1
Calculate the thermal noise voltage generated by a 0.5 MΩ resistor at a temperature of 100°C if the bandwidth is 2 MHz.

We have

$$V_n = \sqrt{(4kTRB)} \text{ volts}$$

where $k = 1.38 \times 10^{-23}$, $T = 100 + 273 = 373K$, $R = 0.5 \times 10^6$ and $B = 2 \times 10^6$. This

$$V_n = \sqrt{(4 \times 1.38 \times 10^{-23} \times 373 \times 0.5 \times 10^6 \times 2 \times 10^6)}$$

$$= \sqrt{(2059 \times 10^{-11})} \text{ volts}$$

$$= 143 \text{ }\mu V$$

A study of equation (3.1) tells us that, to minimize thermal noise, the use of high-value resistances operating at high temperatures must be avoided as far as possible. The temperature of a resistor is not simply that of the ambient but also that resulting from its own generated heat. Strictly, equation (3.1) applies to metallic conductors and so to wire-wound resistors; carbon resistors introduce additional noise because the passage of direct current sets up very small arcs between the carbon granules making up the resistor body. Such internal *contact noise* has no connection with thermally generated noise. The use of low-noise resistors such as metal-oxide types will minimize contact noise.

A noisy resistor can be best represented for purposes of problem solving as a form of Thévenin equivalent generator as shown in *Figure 3.3*. Here the noisy resistance R is replaced by a noise-free ideal resistance of the same value, in series with a voltage generator developing an EMF as given by equation (3.1). The open-circuit voltage across terminals AB is then V_n. No polarity is associated with the hypothetical generator. We can now obtain an expression for the *noise power* developed in a resistance.

Let the equivalent noise generator of Figure 3.3 be connected to a noiseless load resistor whose value is equal to the internal resistance of the source, i.e. $R_L = R$, as shown in *Figure 3.4*.

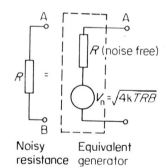

Noisy resistance Equivalent generator

Figure 3.3

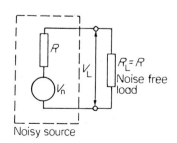

Noisy source

Figure 3.4

This condition, as you will recognize, represents the maximum power transfer from source to load. The voltage across the load will be one-half of the available voltage, that is

$$V_L = \tfrac{1}{2}\sqrt{(4kTRB)} = \tfrac{1}{2}2\sqrt{(kTRB)} = \sqrt{(kTRB)} \text{ volts}$$

Hence the load power is given by

$$P_L = \frac{(\text{load voltage})^2}{\text{load resistance}} = \frac{V_L{}^2}{R_L}$$

Since $R_L = R$,

$$P_L = \frac{kTRB}{R} = kTB \text{ watts} \tag{3.2}$$

Thus the *maximum* power that a thermal noise source can supply to a load when the source and load are matched is kTB watts. Notice that this power is independent of the value of the resistance but is directly proportional to both the temperature and the bandwidth. When the source and load are not matched, equation (3.2) is no longer valid and the noise power transferred to the load is reduced.

Some implications of what is written above may have occurred to you and caused puzzlement. For example, if we connect two resistors in parallel, which one supplies noise power to the other? Well, in the real world, the answer must of course be both. If the circuit containing the noisy resistor R is completed by loading it with another real (i.e. noisy) resistor of any value R_L, then at any given temperature the noise transfer of energy from R to R_L is exactly balanced by the noise energy transfer from R_L to R. Or, put in another way, over a period that is long compared with the average time that an electron travels in a particular direction, the total effective current flow in the circuit is zero. Which is what you would expect from just two resistors connected in parallel!

> 1 Since the transfer of noise power is reduced when the load and source are not matched, would it be advantageous to build an amplifier with deliberately mismatched interstage couplings?

Generators in series

Suppose two noise generators are connected in series. What is the sum total of their equivalent noise output? One thing must be certain: there is no question of us simply adding voltages together *arithmetically*. Noise, remember, is completely random in frequency, phase and instantaneous magnitude. So both generators will set up currents through a load (see *Figure 3.5*) independently of each other and in a completely unrelated (uncorrelated) way. Thus the noise power in the load due to either generator is in no way influenced by the other. So the load noise power is

$$P_n = \frac{V_{n1}^2}{R_L} + \frac{V_{n2}^2}{R_L} = \frac{1}{R_L}\left(V_{n1}^2 + V_{n2}^2\right)$$

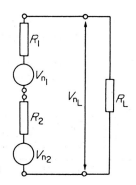

Figure 3.5

and the RMS noise voltage across the load will be

$$V_{nL} = \surd(R_L P_n) = \surd(V_{n1}^2 + V_{n2}^2)$$

Hence noise voltages add as the *square root* of the sum of the squares. In other words, V_{nL} can never be zero, as would be possible if we just added the voltages arithmetically.

Shot noise

Shot, partition and flicker noise are all associated with transistors and valves.

Shot noise in an active device is the result of fluctuations when charge carriers cross potential barriers. In the thermionic valve the noise results from fluctuations in cathode current due to random variations in the cathode space charge. In the transistor, shot noise is caused by the random arrival and departure of carriers by diffusion across a *p-n* junction. A transistor has two such junctions and hence two such sources of noise. Shot noise is most important when low-level signals are being amplified, and in general it is more significant than the thermal noise of associated components.

Partition noise

In a tetrode or pentode valve, the screen grid takes current as well as the anode. There is therefore a division of cathode current between these two electrodes. The screen current has random fluctuations since it is derived from the cathode, and these fluctuations are also superimposed (in the opposite phase) on the anode current.

In the transistor, the input current flows through the emitter to the base–emitter junction, after which it divides between the base and collector circuits, since $I_E = I_B + I_C$. The base current has random fluctuations and these are superimposed on to the collector current. Partition noise is generally negligible in modern transistors, and is not present in triode valves.

Flicker noise

The cause of flicker noise is not fully understood, but it is believed to be due to irregularities in the cathode coating in valves and to emitter surface leakage and recombination in transistors. This noise is predominantly low frequency in nature and becomes insignificant above a few kilohertz. Unlike white noise, flicker (or 'pink') noise departs from the flat relationship with frequency and becomes frequency dependent. For this reason it is often known as $1/f$ noise. Sensitive circuits for low-frequency amplification are particularly affected by the presence of $1/f$ noise, which appears in the FET as well as the bipolar transistor.

Figure 3.6 compares the spectral densities of white noise and pink noise and shows their overall resultant. You should note

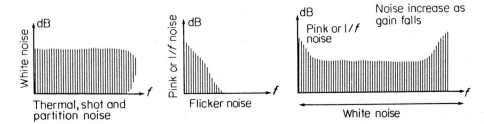

Figure 3.6

that noise increases above the white noise level as higher frequencies are approached because of the fall in gain of the system.

Example 2
Why is a field-effect transistor likely to be less noisy, under the same operating conditions, than a bipolar transistor?

Like the bipolar transistor, the FET exhibits shot noise, thermal noise and $1/f$ noise. However, the FET is inherently a lower-noise device than an ordinary transistor because:
(a) There is only one *p-n* junction involved, so shot noise is reduced.
(b) There is no equivalent base (gate) current and so the source current does not divide before reaching the drain. Partition noise is consequently absent.
(c) Channel resistance is small; hence thermal noise is small.

---

2 If a transistor is operated with a very small collector current, will the shot noise be greater, less or the same as it is when the collector current is large? Give reasons for your answer.

EXTERNAL NOISE Much of the noise originating outside any electronic system is man-made and is, therefore, under control to a certain extent. Other sources contribute natural noise and these are not under control. We deal first with two sources of man-made noise – mains hum and sparks – and then with natural noise.

Mains hum Any electronic system that receives its supply voltage(s) from a rectifying mains power unit, or uses thermionic valves whose heaters are run from low-voltage AC supplies, is liable to have hum introduced into its circuitry. Alternating magnetic flux from wires carrying alternating current, or from the iron cores of transformers and chokes used in the power units, may link with the input circuits of the amplifying devices and induce there an unwanted EMF. Similarly, stray capacitance between active terminals and some other conductor around which there is an alternating electric field will lead to the flow of unwanted current in the active leads.

Electric and magnetic screening, together with a sensible layout and orientation of components will, as mentioned earlier, overcome most of the problems associated with this kind of noise. Hum signals introduced on the actual DC supply leads can be eliminated by additional smoothing and decoupling or by better regulation (stabilization) at the power unit.

Spark or transient interference Whenever an electric circuit is interrupted, such as by a switch, a motor commutator, a car ignition or welding equipment, the resulting sparks generate a high-frequency current in the circuit wires. These in turn act as transmitting aerials, radiating radio signals which cover a very wide band of frequency. There are also voltage pulses set up which pass along the circuit wires in which the switching takes place.

Ideally, when a circuit is interrupted, the current should fall instantly to zero (*Figure 3.7(a)*). However, because of the presence of inductance and capacitance it is possible for a resonant condition to exist which can be shocked into oscillation as soon as the current changes (*Figure 3.7(b)*). The resulting interference can get into an electronic system either by way of the radio-frequency signals being picked up directly on the system wiring or aerial array or by way of the mains input connection.

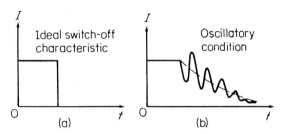

Figure 3.7

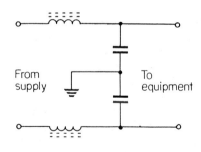

Figure 3.8

Screening the equipment will prevent direct radiation pickup, and choke-capacitance suppressors wired into the mains lead will eliminate most of the voltage pulses coming that way. *Figure 3.8* shows a typical mains suppressor circuit. Aerials cannot be screened, of course, but they can be oriented away from the source of the interference (if this is coming from a fixed direction) or repositioned so that some sort of localized screening is achieved, such as the wall of a house of a line of trees being interposed between the aerial and the source of interference. Raising the aerial and screening the downlead often helps. The apparatus responsible for the interference is the logical place for screening and suppression systems to be fitted.

Ignition system interference is particularly troublesome to car radio receivers and nearby television receivers. Car radio receivers must always be properly earthed to the car chassis and both the aerial and the supply voltage leads must be screened.

Natural noise   The bulk of natural noise is contributed by atmospheric or *static* noise resulting from lightning discharges. The effect of lightning on radio reception is well known, resulting either in violent crackling sounds or, in the presence of sheet lightning, an almost continuous 'frying' background to the wanted programme. Atmospheric noise of this sort can be picked up over very great distances and static interference is often detectable even when no thundery weather is apparent.

At higher frequencies, above some 20 MHz or so, the effect diminishes but other sources of noise begin to show themselves.

Solar disturbances lead to background noise and fading at frequencies above some 50 MHz, and thermal radiation from the surface of the earth, particularly in the summer months, generates noise which becomes prominent at frequencies over 200 MHz. Variations in the ionized layers in the upper atmosphere lead to what is known as quantum noise, though only very high-frequency bands are affected by this. Noise sources in the Milky Way (which are put to good effect in radio astronomy) can be a nuisance to others on frequencies above some 1000 MHz.

SIGNAL-TO-NOISE RATIO

We have already noted that the noise generated in an amplifier or introduced along with the input determines the smallest signal which can usefully be amplified. So important measures of the performance of an amplifier or receiver are what we call the noise factor (sometimes the noise figure) and the signal-to-noise (S/N) ratio.

Signal-to-noise ratio is defined as the ratio of the wanted signal power to the unwanted noise power:

$$\text{S/N ratio} = \frac{\text{signal power}}{\text{noise power}} \tag{3.3}$$

As it is usual to express this ratio in decibels, we have

$$\text{S/N ratio} = 10 \log \frac{\text{signal power}}{\text{noise power}} \text{ dB} \tag{3.4}$$

Since $P = V^2/R$, if we substitute into equation (3.3) for the case where the signal voltage is developed in series with a total resistance R, we get

$$\frac{\text{signal power}}{\text{noise power}} = \frac{(\text{signal voltage})^2}{R} \times \frac{R}{(\text{noise voltage})^2}$$

$$= \left(\frac{\text{signal voltage}}{\text{noise voltage}}\right)^2$$

Therefore

$$\text{S/N ratio} = 10 \log \left(\frac{\text{signal voltage}}{\text{noise voltage}}\right)^2$$

$$= 20 \log \left(\frac{\text{signal voltage}}{\text{noise voltage}}\right) \tag{3.5}$$

Audio amplifiers usually quote signal-to-noise ratio at a frequency of 1 kHz, which then states an average value, but separate figures are sometimes given for both the low- and high-frequency ends of the operational range. Typical figures would be 60 dB for amplifiers and 55 dB for stereo FM radio receivers. In some systems, figures down to 15 dB are acceptable.

Example 3
If the signal level at the input of a radio receiver is 20 μV and the noise level is 1.5 μV, calculate the S/N ratio in dB.

$$\text{S/N ratio} = 20 \log \left(\frac{\text{signal voltage}}{\text{noise voltage}} \right)$$

$$= 20 \log (20/1.5) = 20 \log 13.34 = 22.5 \text{ dB}$$

Now try the following problems on your own.

3 If the signal level at a tape recorder head output is 1 mW and the signal-to-noise ratio is 40 dB, what is the output noise power?
4 If the amplifier following the above tape head has a gain of 35 dB, what will be the S/N ratio at the amplifier output, assuming *no more noise* is introduced?
5 An amplifier has a S/N ratio of power equal to −10 dB at its output when its bandwidth is 10 kHz. What must be the bandwidth so that a +5 dB S/N power ratio is obtained? (Hint: the noise is reduced 15 dB.)

NOISE FACTOR

The measure of the noise quality of an amplifier is the noise factor. The *noise factor F* of an amplifier is the ratio of the total noise at the output to that part of the output noise which is due to the input noise only:

$$\text{noise factor } F = \frac{\text{total noise power at the output}}{\text{output noise power due to input noise power}}$$

It is therefore a measure of the amount of internal noise introduced by the system per unit gain. If the amplifier introduced no noise, F would be unity. Naturally, we should aim to keep F as small as possible.

Figure 3.9 shows an amplifier having a power gain A_p and an internally generated noise power P_{na}. This amplifier is fed from a matched source which provides an input noise power P_{ni}. The total noise power at the output is P_{no}. Then

$$F = \frac{P_{no}}{P_{ni} \times A_p} \tag{3.6}$$

But

$$A_p = \frac{\text{signal power at the output}}{\text{signal power at the input}} = \frac{P_{so}}{P_{si}}$$

Therefore

$$F = \frac{P_{no}}{P_{ni} \times P_{so}/P_{si}} = \frac{P_{si}/P_{ni}}{P_{so}/P_{no}}$$

$$= \frac{\text{input signal-to-noise ratio}}{\text{output signal-to-noise ratio}} \tag{3.7}$$

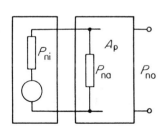

Figure 3.9

Now equation (3.6) can clearly be written in the form

$$F = \frac{P_{no}}{kTB \times A_p}$$

since the maximum power that a thermal noise source can supply to a matched load is kTB watts (from equation (3.2)). Then $P_{no} = FkTBA_p$. This output noise power is made up of amplified input noise $(kTBA_p)$ and amplified internal noise, so that the internal noise power P_{na} is the difference given by $P_{no} - kTBA_p$:

$$P_{na} = FkTBA_p - kTBA_p$$

$$= (F - 1)kTBA_p \qquad (3.8)$$

Let the amplifier shown in *Figure 3.10* be made up of an input stage having a noise factor F_1 and power gain A_{p1} and the subsequent stage or stages having a noise factor F_2 and power gain A_{p2}. We assume that the useful bandwidth B is limited by the subsequent stages. The input provides a noise power kTB watts, and after amplification this appears at the output as $kTB\ A_{p1}A_{p2}$ watts.

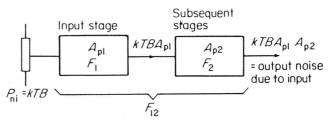

Figure 3.10

From equation (3.8) the internal noise from the input stage is $(F_1 - 1)kTBA_{p1}$ and this appears at the output as $(F_1 - 1)kTBA_{p1}A_{p2}$. Now the *overall* noise power originating from the input is

$$F_{12} = \frac{\text{total noise output}}{\text{noise at output due to input}} = \frac{\text{total noise output}}{kTBA_{p1}A_{p2}}$$

$$= \frac{kTBA_{p1}A_{p2} + (F_1 - 1)kTBA_{p1}A_{p2} + (F_2 - 1)kTBA_{p2}}{kTBA_{p1}A_{p2}}$$

$$F_{12} = F_1 + \frac{F_2 - 1}{A_{p1}} \qquad (3.9)$$

A study of this last equation reveals that the overall noise factor of a cascaded amplifier is primarily influenced by the noise introduced in the first stage. To make F_{12} low, F_1 must be low and the first-stage gain A_{p1} must be very large.

Example 4
An aerial has a noise power equal to 0.1 μW and the noise factor of a receiver matched to this aerial is 10. Calculate (a) the total noise power at the output of the receiver if it

has a power gain of 80 times (b) the proportion of the output noise power contributed by the receiver.

(a) From equation (3.6) the output noise power is given by

$$P_{no} = FP_{ni}A_p = 10 \times 0.1 \times 80 = 80 \ \mu W$$

(b) The noise power at the output due to the input noise is $0.1 \times 80 = 8 \ \mu W$. Thus the amplifier contributes $80 - 8 = 72 \ \mu W$ of noise, which is 90 per cent of total noise.

---

6 A receiver has a power gain A_p, a bandwidth B Hz and a noise factor F. Prove that the noise power introduced by the receiver can be expressed as $(F - 1)kTBA_p$ watts.

Bear in mind that although we have so far expressed noise factor F as a numerical ratio, it can be expressed in dB in the usual way. A typical figure for a communications receiver would be 6 dB. A figure of 3 to 4 dB would be considered very good.

---

7 Give an example of (a) a low-frequency noise (b) a high-frequency noise (c) a man-made noise (d) a natural noise.

8 What is the noise voltage of a 100 kΩ resistor at 27°C if the bandwidth is 1 MHz?

9 If the signal level at the input to a receiver is 20 μV and the noise level is 1 μV, what is the signal-to-noise ratio in dB?

10 The signal level from a microphone output is 5 mW. If the S/N ratio is 40 dB, what is the output noise power?

11 Why is an FET less noisy than a bipolar transistor?

12 What is meant by the terms white noise, pink noise and $1/f$ noise?

An amplifier has a flat passband characteristic extending from 100 kHz to 500 kHz and a gain of 40 dB. What will be the noise voltage at its output resulting from a 500 kΩ resistor at 20°C connected across its input terminals?

13 If the effective noise level at the input to a receiver is −130 dB relative to 1 mW and the signal-to-noise ratio expected is 23 dB, what receiver gain is required to produce an output of 25 mW? (*Note*: a dB level referred to 1 mW is symbolized as dBm: see Section 1.)

(C & G)

14 A receiver is receiving a signal, and measurements show that the S/N ratio at the receiver output is 30 dB. Assuming that all the noise occurs in the first stage of the receiver, calculate the output S/N ratio when (a) the output stage gain of the receiver is increased by

6 dB (b) the incoming signal fades so that the receiver power falls to one-quarter of its previous value.

15 Prove that if an amplifier has a gain G and a noise factor F, then the product $G(F-1)$ is proportional to the internal noise.

16 What is meant by the noise factor of an amplifier? Prove that the power delivered to a matched load by a noise source of internal resistance R is kTB watts.

A communications receiver has a noise factor of 10 dB at 30 MHz. Is this a poor, a typical or a very good figure?

17 If the noise power in a given bandwidth in an aerial is 25 pW (1 pW = 10^{-12}W) and the noise factor of a receiver matched to the aerial is 10, calculate (a) the total noise power in the given bandwidth at the output of the receiver if it has a power gain of 50 (b) the noise power at the output due to the input noise (c) the contribution of the amplifier to the output noise power. What would happen, if anything, to the noise power if the bandwidth was halved?

18 An amplifier has a gain of 20 dB and a 6 dB noise factor. An aerial produces an available thermal noise power in a given bandwidth of -110 dBm. If the aerial is attached to the amplifier input and both are at the same temperature, in a given bandwidth what is (a) the total noise power at the output (b) the noise power at the output due to the amplifier noise only? If an identical amplifier is cascaded with the first, what will be the noise factor of the combination?

4 Oscillators

Aims: At the end of this unit section you should be able to:
Understand the operation of an oscillator as an amplifier with positive feedback.
Explain the operation of LC tuned oscillators and RC oscillators as sinusoidal oscillators.
State the factors which affect frequency stability.
Explain the operation of a crystal-controlled oscillator and state its advantages.
Construct and test a Wien bridge oscillator using an operational amplifier.

Oscillators are extremely important elements in electronic and communications engineering. They provide signal sources for electronic measurement and are generally used as such in combination with other measuring devices under the name of signal generators or function generators. Oscillators operating at fixed frequencies, or operating only over a restricted range of frequencies, are also to be found in transmitters and in all radio and television receivers.

An oscillator is essentially a power converter, in the sense that its only input is the DC power supply and its output is a continuous waveform which may or may not be sinusoidal in shape. A voltage amplifier will, in general, convert a small alternating input signal into a large alternating output signal. If a part of the output is fed back to the input of the amplifier, then the amplifier will be providing its own input. In the circuit diagram of an oscillator, therefore, we will not expect to find a terminal which is specifically the input signal terminal, though there will be an output terminal from which the generated oscillation may be derived.

There are two basic forms of oscillator:

1 Those in which the generated waveform is basically sinusoidal. These are known as sinusoidal or harmonic oscillators and generally take the form of tuned-feedback or negative-resistance oscillators;
2 Those in which the generated waveform is markedly non-sinusoidal, being characterized in fact by abrupt changes from one condition of circuit stability to another. These oscillators are known as relaxation oscillators.

Our concern in this section is to examine examples of both these types of oscillator.

CRITERION OF OSCILLATION
From Section 2 we recall the expression for the overall gain A' of an amplifier with feedback:

$$A' = \frac{A}{1 - \beta A}$$

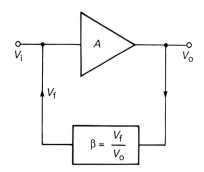

Figure 4.1

where A is the gain of the amplifier without feedback, and β is the feedback ratio V_f/V_o (see *Figure 4.1*). We recall also that mention was made of the particular condition which comes about when the denominator of this expression equals zero, for then the loop gain $\beta A = 1$ and A' becomes infinite. This condition represents oscillation, for it implies that an output signal is available when the external input is zero; and this relates to our description of an oscillatory system given above, where a continuous signal output is obtained without any form of *external* signal input terminal being apparent.

There are two conditions which must be satisfied if such a system is to work properly, and we can consider these in relation to Figure 4.1:

1 The amplifier must be capable of supplying sufficient power to compensate for that which is dissipated in the resistance of the feedback network, and still provide enough power to its own input terminal to maintain the required level of useful output.
2 There must be an overall zero (or 360°) phase shift round the complete loop so that the output signal from the feedback network will be precisely in step with the required amplifier input signal. Hence, if we were using a single-stage common-emitter amplifier, for example, which was itself introducing a 180° phase shift between input and output, there would have to be a further 180° phase shift in the feedback network if oscillation was to be possible.

We can deduce from a consideration of these two criteria that the amplifier circuit forms that part of the system which maintains the *amplitude* of the output waveform, and it is the feedback network which determines the *frequency* of the generated waveform. In practice, the loop gain βA must actually be greater than unity; for if it were precisely unity, then variations in the supply voltages, the temperature and other environmental factors would in all likelihood result in a collapse in the output because of a reduction in the overall gain.

When βA is made greater than unity (and β is made positive) more signal is fed back than is actually required for oscillation to occur, and a buildup in signal level around the loop follows. This is a necessary condition for oscillation to get going. Such a buildup cannot continue indefinitely, but is quickly limited by non-linearities within the amplifier and by the finite value of the supply voltage. Such non-linearities, as we have noted earlier, will always introduce distortions into the output waveform, but if βA is kept very close to unity and always positive then such distortions can be held to negligible proportions.

TUNED SINUSOIDAL OSCILLATORS

The most elementary, and often the most common, sinusoidal oscillators are those which comprise a transistor and a tuned circuit. The tuned circuit – capacitance in parallel with inductance – may be connected into the base circuit of the amplifier (the tuned-base oscillator) or into the collector circuit (the tuned-collector oscillator). In both cases, the transistor amplifies at the resonant frequency of the tuned circuit and the collector voltage is in phase opposition to the base voltage. A fraction β of the collector voltage is taken, reversed in phase by an appropriate coupling connection and fed to the base circuit. The transistor thus provides its own input and the system oscillates because the total phase shift is 360°.

In such circuits, the frequency of oscillation is determined by the *LC* product and, to a close approximation,

$$f = \frac{1}{2\pi\sqrt{(LC)}} \tag{4.1}$$

This is the resonant frequency of a high-Q parallel-tuned circuit.

Figure 4.2(a) shows the tuned circuit L_1C in its position as the frequency-determining element in the feedback path of a typical feedback system. Suppose that amplifier A is a single-stage common-emitter amplifier. A phase shift of 180° occurs within the amplifier and, by the correct arrangement of the coupling coil L_2 relative to L_1, a further 180° shift will occur within the feedback. A total shift of 360° is now accomplished.

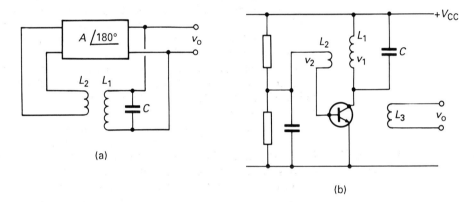

(a)

(b)

Figure 4.2

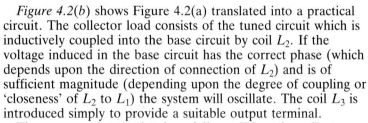

Figure 4.2(b) shows Figure 4.2(a) translated into a practical circuit. The collector load consists of the tuned circuit which is inductively coupled into the base circuit by coil L_2. If the voltage induced in the base circuit has the correct phase (which depends upon the direction of connection of L_2) and is of sufficient magnitude (depending upon the degree of coupling or 'closeness' of L_2 to L_1) the system will oscillate. The coil L_3 is introduced simply to provide a suitable output terminal.

The process of operation is as follows. When the collector supply is switched on, collector current starts to flow and the increasing magnetic flux around L_1 links with the turns of L_2 and induces an EMF in the base circuit; this EMF will develop between base and emitter as an input signal. A simplified phasor diagram is shown in *Figure 4.3*. Starting with collector current i_C, the voltage v_1 across the tuned circuit will be in phase with i_C (since the circuit is resonant at its natural frequency and hence purely resistive). The current i_L through L_1 will lag on i_C by almost 90°, and the EMF v_2 induced in the base circuit will lead or lag on i_L by 90° depending upon the sign of

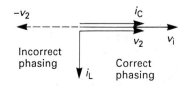

Figure 4.3

the coupling between the coils; this in turn depends simply upon the direction of connection of coil L_2. This induced EMF acts as an input voltage (v_2) and, if this is in phase with collector current i_C, energy is being fed back in such a manner that the circuit is self-exciting.

There are a number of possible circuit arrangements involving a transistor amplifier and a resonant LC circuit in the feedback path, and you will encounter some of these in your practical work. In nearly all of them the frequency will be controlled by C. If the oscillator is going to be used at a single frequency it is usual to employ a fixed, stable capacitor for C and to adjust L by means of a ferrite core; this core is sealed in position after the frequency has been set. In oscillators required to cover a wide range of frequency, the value of L may also be made variable by switching in different coils to provide an overlapping set of frequency ranges, each tuned by a single capacitor.

The LC tuned circuit type of oscillator can be used to generate sinusoidal waveforms up to many tens of megahertz without difficulty. However, for low-frequency operation (of the order of a few hundred hertz and less) the values of both L and C become large, and resistive losses (as well as physical bulk), particularly in the inductor, become prohibitive. It is then preferable to use resistance-capacitance feedback networks and the oscillator forms which go under the general name of RC or *phase-shift* oscillators. The output waveform from this type of oscillator is also sinusoidal, in contrast to *relaxation* oscillators which, while using R and C in their feedback paths, generate non-sinusoidal waves.

CRYSTAL-CONTROLLED OSCILLATORS

The frequency stability of circuits similar to the tuned collector already discussed will not normally be better than about ten parts in a million per °C temperature variation. This figure can be improved considerably by replacing the tuned circuit by a piezoelectric crystal. Rochelle salt, tourmaline and quartz crystals all exhibit the piezoelectric effect to a marked degree. If pressure is exerted between opposite faces of a thin slice of such crystals, a potential difference is set up between the faces. If the pressure is changed to a tension, a potential difference of opposite polarity is created. If alternating cycles of pressure and tension are applied to the faces of the crystal, therefore, an alternating voltage is established across the faces. The converse effect can also be produced; if an alternating voltage is applied to the crystal, a mechanical vibration tends to occur.

The amplitude of this vibration is very marked at the natural frequency (or mechanical resonance) of the crystal slice, and in this respect the system resembles a tuned LC circuit. The change in amplitude of the vibrations for a small change in the frequency is very large, and a crystal used in this way can readily exhibit an equivalent Q-factor of several thousands.

The natural frequency of a crystal slice depends on a number of factors, of which the chief are:

1 The type of crystal, though generally quartz is used for radio frequencies.

2 The crystal cut, that is the inclination of the slice in relation to the various axes of the crystal from which it is cut.
3 The dimensions of the slice; the frequency increases as the slice becomes thinner.
4 Temperature, although special cuts can be chosen which give an almost zero temperature coefficient. For precise control the crystal may be mounted in an 'oven' enclosure, the temperature of which is held just above ambient within very close limits.

In addition to the above, the crystal frequency also depends to some extent upon the method of mounting it into a holder. Contacts are often sputtered directly on to the crystal faces or parts of the faces, and leads are brought out from these. Alternatively the crystal slice is placed between a pair of flat, parallel metal electrodes. In this method a very thin air gap is left between the crystal and the electrodes, and a small frequency adjustment can be introduced at the manufacturing stage by a variation in the thickness of this gap (see *Figure 4.4*). This facility is not, of course, available to the user, but a small-value capacitor in parallel with the crystal is often introduced as an adjustment aid. Crystals are readily available with fundamental frequencies from some 5 kHz up to about 15 MHz. Higher frequencies are catered for by operating crystals at a harmonic of their fundamental frequencies. These are known as *overtone* crystals, and they are available in frequencies up to some 100 MHz. It is easily possible to detect harmonics up to thousands of megahertz from such overtone crystals, and some frequency standards are based on this technique.

The equivalent circuit of a crystal is illustrated in *Figure 4.5*, and the consequent behaviour is that both series resonance and parallel resonance are exhibited at frequencies f_s and f_p respectively. Parallel resonance always occurs at a slightly higher frequency than series resonance, the separation being of the order of 0.5 per cent of f_s, but for most practical purposes the two are considered to be coincident. Because of this effect, it becomes possible to use a crystal in a feedback loop to control the frequency of oscillation of a number of different forms of oscillator.

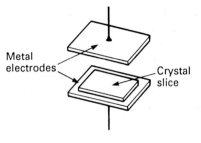

Metal electrodes

Crystal slice

Figure 4.4

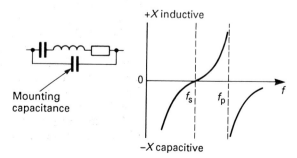

Mounting capacitance

+X inductive

−X capacitive

Figure 4.5

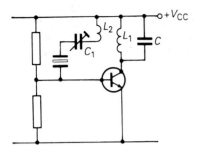

Figure 4.6

Figure 4.6 shows one method in which the crystal is connected in the feedback coupling of the circuit discussed above. The crystal is operated in its series-resonant mode so that it has a minimum impedance at the required oscillation frequency. Capacitor C_1 provides a fine adjustment of this frequency. We have, in effect, a crystal-controlled tuned-collector oscillator. Circuits of this sort can be used up to about 100 kHz in frequency.

Not all crystal oscillators need to use discrete transistors in this way. A little later on we will look at oscillators using operational amplifiers, but it is common practice nowadays to utilize both TTL (7400 series) and CMOS (4000 series) logical gate systems as oscillators. *Figure 4.7* shows a typical circuit which can utilize crystals of 1 MHz frequency and higher in this way, particularly for the generation of precise clocking and control signals after suitable squaring. Here three of the four available two-input NAND gates from a 7400 package are used.

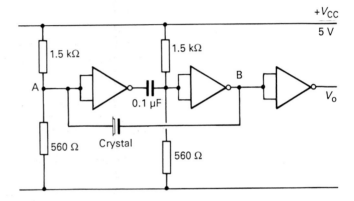

Figure 4.7

By strapping the twin inputs of each gate together, the gates become inverters. And by biasing the inverters by means of resistor dividers, the greatest gain is obtained from each of them. Since each inverter introduces a 180° phase shift, the output at point B is in phase with the input at point A, and at its series resonance frequency the crystal, which is in the feedback path, provides the in-phase feedback condition to produce oscillation. The third section of the 7400 is used as a simple buffer stage to isolate the oscillator proper from later circuitry.

Example 1

The phase shift between the input and the output terminals of a certain feedback network is given by

$$\phi = \frac{270°}{1 + f/500}$$

where f is in hertz. This network is used in an oscillator circuit where a single common-emitter amplifier is employed. At what frequency will this circuit oscillate?

Since the amplifier itself will introduce a 180° phase shift, the feedback network must itself introduce a further 180′ shift. Hence the phase angle φ must be 180°. From the given expression, therefore,

$$1 + \frac{f}{500} = \frac{270}{180} = \frac{3}{2}$$

$$\frac{f}{500} = \frac{1}{2}$$

$$f = 250 \text{ Hz}$$

Strictly we should work in radian measure instead of degrees, but the same answer will be obtained.

Example 2

Why does the stability of an oscillator using resonant tuned circuits improve as the Q-factor of the tuned circuit is increased?

A high Q-factor resonant circuit has a large rate of change of phase angle with frequency close to resonance. Thus the higher the Q-factor, the smaller will be the change in frequency brought about by any alteration in the phase angle within the *amplifier*. We have already noted that it is the phase relationships around the feedback loop which determines the oscillation frequency. If a small phase shift occurs in the amplifier, the oscillation frequency will change to maintain a total shift of zero around the feedback loop. Circuits which require the greatest frequency stability, therefore, must use resonant circuits of the highest possible Q-factor. This is where crystal-controlled circuits having Q-factors of hundreds or thousands have the advantage.

RESISTANCE-CAPACITANCE OSCILLATORS

It is not necessary to employ resonant LC circuits in the feedback paths of oscillators, and various arrangements of resistance and capacitance can be used instead. It might seem on first consideration that a network made up of resistors and capacitors would not only be non-resonant (in the sense that LC circuits are resonant) but would also introduce a considerable resistive loss. Both of these points are valid; a resistor-capacitor network cannot exhibit resonance in the usual meaning of that term, and there is bound to be resistive loss. But we must keep in mind that we are concerned with *phase shift* in the feedback network, not resonance; and resistive loss can always be made good in the amplifier part of the loop.

We recall from basic theory that when an alternating voltage is applied to a series RC circuit, as in *Figure 4.8(a)*, the phasor diagram will be as in *Figure 4.8(b)*. The voltage across the resistor, $V_R = IR$, will lead the applied voltage V by angle ϕ. The output voltage V_o across the capacitor, $V_C = IX_C$, will lag V by an angle $(90 - \phi)$. This circuit can therefore be used to change the phase of a signal. Because the output lags the input, this is a *phase-retard* circuit. If C and R are interchanged so that the output is derived across R, the circuit becomes a *phase-advance* circuit. Either of these arrangements may be used in phase-shift oscillator systems.

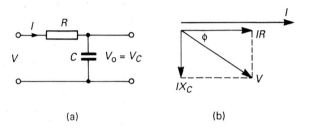

(a)   (b)

Figure 4.8

> 1 What is the greatest phase shift possible with these circuits? What happens to the output voltage as the phase shift increases?

If you have attempted both parts of this problem (and checked the given solutions), you may have realized the serious disadvantage of these circuits. The greatest phase angle possible is clearly 90°, but under this condition the output voltage from either connection is zero. Or, if you like, the attenuation is infinite.

Suppose we are using an amplifier with a phase shift of 180° in a proposed oscillator circuit. Then a further 180° shift is required in the feedback network. One RC network of the form just discussed will provide a maximum shift of 90° only; two such networks in cascade will provide a maximum shift of 180° but with infinite attenuation. We shall require three networks at least to ensure that we can get a 180° shift with moderate attenuation. Each section of the so-called *ladder* network will then introduce a 60° shift at one particular frequency.

Figure 4.9 shows how a phase-advance network can be connected between the output and the input of a common-emitter amplifier. If the loop gain at a particular frequency is greater than unity with a total phase shift of 360°, oscillation will take place. Think about the network in this way: at low frequencies the series capacitors will have very high reactances, so that the loop gain will be low and overall will be less than unity. At very high frequencies the capacitors will behave as virtual short-circuits but the phase shift will be small. There will be one frequency only at which the loop gain and the phase shift will be correct for oscillation. It can

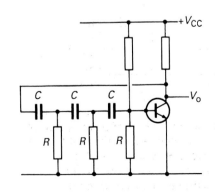

Figure 4.9

be demonstrated that this frequency, at which the ladder phase shift is 180°, is given by

$$f = \frac{1}{2\pi\sqrt{(6)}CR}$$

and that at this frequency the ratio $V_o/V_i = \beta = -1/29$. Hence a gain of at least 29 is necessary from the amplifier to provide a loop gain of unity.

It might seem that it would be necessary to ensure that each section of the ladder network was precisely identical, so calling for the use of precision components. This is not so, and it is sufficient to use standard 5 per cent tolerance parts. This leads to a modification of the attenuation within the network to a figure greater than 1/29 somewhat, but the generated frequency will adjust itself in such a way that the total ladder phase shift adds up to 180°, though not in three identical 60° shifts. In any event, the loading effects of the transistor input and output terminals affect the generated frequency in some respects.

Phase-shift oscillators of this sort are particularly useful where a fixed output is required at a relatively low frequency. It is difficult to make these oscillators tunable over a range of frequencies, as either all three capacitors or all three resistors have to be made simultaneously adjustable.

An Experimental Circuit A phase-shift oscillator using an operational amplifier is shown in *Figure 4.10*. The feedback connection to the amplifier is taken to the inverting input so that a 180° phase shift takes place within the amplifier. The remaining 180° takes place in the frequency selective ladder made up of three capacitors and three resistors. As the required gain in the amplifier is (theoretically) 29, the value of resistor R_f should be at least 28 times the value of R_1. To make sure that the oscillation will get going, this resistor is actually made about 32 times greater than R_1.

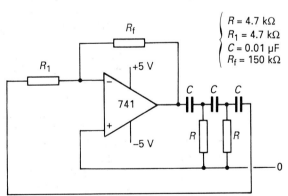

Figure 4.10

Alternatively, in an experimental setup, it is useful to replace R_f with a potentiometer of value 250 kΩ and adjust this until oscillation just begins and the waveform is free of distortion.

The output frequency should be measured using a digital frequency meter and compared with the theoretical value obtained from the formula given earlier.

THE WIEN BRIDGE OSCILLATOR

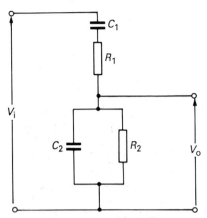

Figure 4.11

This is another example of an oscillator using a resistor-capacitance network in the feedback loop. Unlike the phase-shift oscillator, however, the network introduces a zero phase shift, and the amplifier consequently requires a phase shift which is also zero.

Any network made up of pure resistance will have a zero phase shift at all frequencies, but by introducing capacitors and making up a network as shown in *Figure 4.11*, zero phase shift will be obtained at only one frequency. This network is known as the Wien or frequency bridge because it forms a part of the Wien alternating-current bridge used for the measurement of frequency. It is usual to make $R_1 = R_2$ and $C_1 = C_2$ in this network. The frequency at which zero phase shift occurs between input and output is then given by

$$f = \frac{1}{2\pi CR} \tag{4.2}$$

where $C = C_1 = C_2$ and $R = R_1 = R_2$.

If the network is used in the feedback path of an amplifier which has itself zero (or 360°) phase shift, then oscillation should occur at the frequency dictated by the bridge component values. As the network will obviously introduce an attenuation, an equivalent or greater gain will be required from the amplifier.

It can be shown that the network introduces an attenuation of 3; that is, the ratio $V_o/V_i = 1/3$. A gain of at least 3 is therefore called for from the amplifier part of the circuit. There is no problem in obtaining such a small gain as this, and, if two common-emitter amplifiers are used (to obtain a total phase shift of 360°), the very high gain resulting from such a circuit enables a considerable amount of negative feedback to be used so that the gain is reduced to the required level with assured gain stability.

The Wien oscillator has several advantages over the phase-shift and LC oscillators discussed above. The oscillation frequency is easily adjusted by using either ganged resistors or capacitors; the circuit operates down to very low frequencies with excellent waveform; and, for a given variation in C, it provides a much wider frequency variation than do circuits using inductance and capacitance, since f varies as $1/C$ and not $1/\sqrt{C}$ as in oscillators using LC feedback. In fact, it is in the elimination of large inductances that the Wien oscillator is such a popular audio-frequency generator. In practical designs it is usual to make R variable by using a two-gang potentiometer and having a switched selection of fixed capacitors to provide decade ranges of frequency, though C can be made the variable.

A typical design system using what is sometimes known as a 'ring of three' configuration is shown in *Figure 4.12*. It may seem odd at first that three transistors (an FET and two bipolars) are used when two would do. However, the form the amplifier actually takes depends not only upon its gain and phase-shift characteristics, but also upon its input and output resistance, both of which affect the Wien feedback network operation. From a consideration of the formula for the

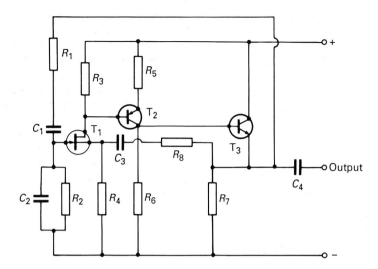

Figure 4.12

frequency of zero phase shift in the network, the generation of low frequencies below, say, some 20 Hz or so calls for quite large values of resistance – assuming that C is composed of a twin-gang 500 pF variable capacitor for tuning purposes, as it sometimes is. For example, for $C = 500$ pF and a frequency of 20 Hz, we need

$$R = \frac{1}{2\pi fC} = \frac{10^{12}}{2\pi \times 20 \times 500} \text{ ohms}$$

$$\approx 16 \text{ M}\Omega$$

As the lower part of the Wien network is effectively shunted by the input resistance of the amplifier, it is necessary to make the input resistance very high in relation to the Wien resistance. The FET, used in the common-source mode, has such a high input resistance. It also introduces the first 180° phase shift.

Transistor T_2 is used as a directly coupled common-emitter amplifier and so introduces a further 180° shift. This in turn feeds into T_3, which is an emitter-follower (common-collector) configuration; there is no phase shift in this stage and hence the overall amplifier phase shift is 360° (or zero) as required. The purpose of the emitter follower is twofold: to provide a low output resistance which matches into the Wien network from this end, and to act as a buffer against the influence of loading on the output terminals. Negative feedback is applied to the amplifier by way of R_8, C_3 and R_4. This reduces the gain to a level which is stable so that oscillations can be maintained without overloading and distortion.

Some More Experiments

It is instructive to make up a Wien-type oscillator in two forms, one using an opamp and the other using discrete transistors. The type using an opamp is usually much easier to assemble on protoboard or similar, and an experimental circuit is given in *Figure 4.13*. The feedback is taken to the non-inverting input of the amplifier so that there is zero phase shift, and negative

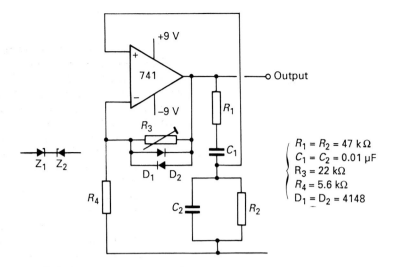

$R_1 = R_2 = 47 \text{ k}\Omega$
$C_1 = C_2 = 0.01 \text{ μF}$
$R_3 = 22 \text{ k}\Omega$
$R_4 = 5.6 \text{ k}\Omega$
$D_1 = D_2 = 4148$

Figure 4.13

feedback is applied to the inverting input to reduce the overall gain to about 3. For this it is necessary for R_3 to be made equal to $2R_4$ since the gain is given by $(R_3/R_4) + 1$. Hence R_3 is made adjustable to get oscillations going.

Make up this circuit using a 741 opamp. Carefully adjust the feedback resistor R_3 until a stable undistorted sine-wave output is obtained on the CRO. Measure the amplitude of this output and use a suitable frequency meter to find its frequency. Compare your frequency measurement with a calculation made from the formula (4.2) earlier. Account for any discrepancy. Adjust (if you can) the V_s supply to the amplifier, making it, say, first ±5 V and then ±12 V. Do these changes have any effect on either the frequency or the amplitude of the output? If they do, explain why. With the component disconnected at one end, measure the resistance of R_3 at the setting used for the proper output from the oscillator. Record this value.

Now replace the two parallel diodes D_1 and D_2 with two series-connected 3.6 V zener diodes as shown at the side of Figure 4.13. Repeat the above procedures in turn, making a note of any difference between the results obtained now with those obtained previously. You might, if time permits, try replacing R_1 and R_2 with a two-gang 10 kΩ linear potentiometer and investigate the range of frequencies you can get by adjusting this. Is the output amplitude constant over the range? Does the waveform remain undistorted?

You will have wondered about the purpose of the diodes or the zeners in this circuit. The circuit will, in fact, work without them provided the ratio of R_3 to R_4 is set to be equal to 2. Try it, and afterwards measure R_3 again. Does it fit the calculated value? And how does it compare with the earlier value you measured when the diodes were initially in parallel with it?

The purpose of the diodes is to stabilize the amplitude of the oscillator output. This is often done by using a thermistor (or a low-current lamp bulb) in place of the feedback resistor, but the principle is similar: to adjust the feedback so as to compensate for variations in the output. When the output amplitude has settled down, the AC feedback voltage across R_3 will cause both diodes to provide some forward conduction on both half-cycles, and this, in conjunction with the set value of R_3, constitutes the *effective* feedback resistance. If now, say, the output tends to rise, the diodes will conduct more heavily, reduce this effective resistance, and so reduce the amplifier gain correspondingly. The output will then be restored. Conversely, when the output tends to fall, the amplifier gain is slightly increased.

You should be able to deduce for yourself how the zener diodes perform the same sort of service. Which do you think is most effective?

A Wien oscillator using discrete transistors is given in *Figure 4.14*. (This is based on an ITT application note, to which acknowledgement is made.) Transistors T_1 and T_2 form a high-gain complementary pair with an overall phase shift of zero and a high input impedance. The familiar Wien network is seen feeding from output to input. Set up this circuit and adjust R_6 (very carefully) to get an undistorted but stable sine-wave output. Repeat the procedures outlined for the previous experiment. This circuit is specifically designed for the generation of low frequencies.

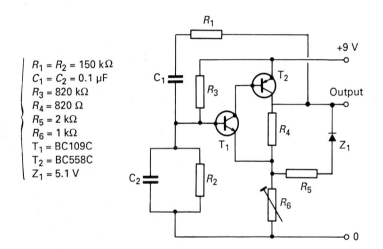

$R_1 = R_2 = 150\ \text{k}\Omega$
$C_1 = C_2 = 0.1\ \mu\text{F}$
$R_3 = 820\ \text{k}\Omega$
$R_4 = 820\ \Omega$
$R_5 = 2\ \text{k}\Omega$
$R_6 = 1\ \text{k}\Omega$
$T_1 = \text{BC109C}$
$T_2 = \text{BC558C}$
$Z_1 = 5.1\ \text{V}$

Figure 4.14

The zener diode provides a stabilizing influence on the output amplitude, though only the positive half-cycle, now cause it to pass its voltage threshold, after which R_4 is progressively shunted by R_5 and the output is reduced. Try the effect of other zener voltage values between 2.7 V and 4.7 V; you may need to adjust R_6 each time you change things.

Example 3
A Wien-type network is designed for a zero phase shift at a frequency of 10 kHz. If the capacitor values used are 7500 pF, what values are required for the resistors?

The frequency at which zero phase shift occurs is given by

$$f = \frac{1}{2\pi CR}$$

from which

$$R = \frac{1}{2\pi fC} = \frac{10^{12}}{2\pi \times 10^4 \times 7500} = 2122 \ \Omega$$

It would be necessary to choose preferred 2200 Ω resistors in this case.

Example 4
The network of the previous example is connected to an amplifier that has an output resistance of 500 Ω and an input capacitance of 500 pF. At what frequency would oscillations actually occur with this circuit?

We are using a feedback network in which $C_1 = C_2 = 7500$ pF, $R_1 = R_2 = 2200 \ \Omega$. When this is connected to the amplifier, the *effective* component values of the Wien network become as shown in *Figure 4.15*. The output resistance of 500 Ω appears in series with 2200 Ω, and the input capacitance of 500 pF appears in parallel with 7500 pF.

The frequency for zero phase shift is now

$$f = \frac{1}{2\pi\sqrt{(C_1C_2R_1R_2)}}$$

$$= \frac{1}{2\pi\sqrt{(8000 \times 10^{-12} \times 7500 \times 10^{-12} \times 220 \times 2700)}}$$

$$= 8430 \text{ Hz}$$

Notice how the input and output impedances of the amplifier can alter the generated frequency from that predicted by theory.

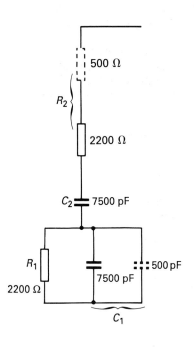

Figure 4.15

---

2 What components may be used as frequency-determining elements in an oscillator?

3 Explain why an oscillator may be regarded as an amplifier with positive feedback.

4 A tuned-collector oscillator circuit (which has no faulty components whatsoever) fails to oscillate when switched on. Where would you first look for the possible cause of the trouble?

5 A tuned oscillator has $L = 15$ mH, $C = 500$pF. At what frequency will this oscillator operate?

6 What distinguishes a harmonic oscillator from a relaxation oscillator?

7 The criterion for oscillation in an ideal system is a loop gain of unity and zero phase shift. True or false?

8 A student answering an examination question wrote: 'An oscillator is an unstable amplifier.' Comment on the aptness, or otherwise, of this statement.

9 Name two oscillator systems in which the amplifier and the feedback circuit both produce a phase shift of 180°.

10 What are the limits of phase advance and phase retard in circuits using a single resistor and a single capacitor? Sketch phasor diagrams for angles of phase shift approaching 90°. What can you deduce from these diagrams?

11 Draw the circuit diagram of a Wien bridge oscillator using either discrete transistors or an integrated circuit amplifier. Identify the components which (a) determine the frequency (b) determine the amplitude of the output.

12 *Figure 4.16* shows a tuned *LC* oscillator which is known as a tuned-base oscillator. Deduce how the circuit operates. Can you see a possible disadvantage this circuit might have compared with the tuned-collector oscillator?

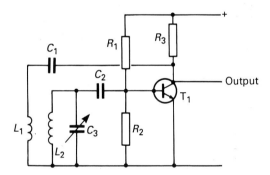

Figure 4.16

13 The components used in the feedback network of a Wien oscillator similar to that shown in Figure 4.13 are $C_1 = C_2 = 10\ \mu\text{F}$, $R_1 = R_2 = 4.7\ \text{k}\Omega$. Estimate the frequency of oscillation. Why is your calculation likely to be only approximate?

14 A certain transistor amplifier has a phase shift of 360° and effective input and output resistances of 1.2 kΩ and 2.5 kΩ respectively. It is used as a Wien oscillator with the feedback component values being $R_1 = R_2 = 15\ \text{k}\Omega$, $C_1 = C_2 = 0.004\ \mu\text{F}$. Obtain estimations for (a) the frequency of oscillation (b) the gain of the amplifier when the oscillations are stable. (Note: the answer to part (b) is *not* 3.)

5 Operational amplifiers

Aims: At the end of this unit section you should be able to:
Be familiar with what an operational amplifier is.
Compare the ideal operational amplifier with a practical device.
Understand the working of operational amplifiers in various circuit configurations.
Design and test a simple operational amplifier using manufacturers' data.

Not so many years ago the operational amplifier (opamp) was a specialized and rather expensive piece of equipment found only in research establishments. Its original role was, and still is in many applications, to perform the mathematical operations of multiplication, division, differentiation and integration, particularly in applications to analogue computing. The early systems used thermionic valves and were of necessity very bulky articles, besides having serious problems of heat dissipation and ventilation. With the arrival of the transistor, such problems disappeared. The transistor's advantages of small bulk, low power consumption and reliability soon brought the opamp out of the exclusive laboratory atmosphere and into an almost limitless variety of everyday electronic systems. The basic operational amplifier as now available in integrated circuit form is a high-performance, directly coupled amplifier capable of high gain and stable operation over a wide range of frequencies including DC.

THE IDEAL AMPLIFIER

The circuit analysis of the opamp can be reduced for our immediate purposes to a few simple restraints. It is not necessary to know the full internal circuit details of the amplifiers in order to use them (such analyses can often be traumatic!) but it is necessary to be aware of the facilities provided by them and the terms used to specify their performance.

As we have mentioned, the opamp is a high-gain, DC coupled system with a frequency range extending down to zero. The output is normally single ended but the input can be either single or double (differential) ended. For single-ended applications, one of the differential inputs is earthed.

An ideal model of the opamp would have the following characteristics:

1 Infinite voltage gain
2 Infinite input impedance
3 Infinite bandwidth starting at DC
4 Zero output impedance
5 Zero output when the differential inputs are identical.

No practical designs can achieve this ideal, of course, but they can approach it. The very common 741 opamp, for example, which is obtainable for many general purpose designs, meets the requirements in this way:

Voltage gain   > 100 000
Input resistance ≃ 2 MΩ
Bandwidth   ≃ 0 to 1 MHz
Output resistance ≃ 75 Ω

When the ideal characteristics are considered, the circuit model reduces to that shown in *Figure 5.1(a)*. Here two input signals are applied with respect to earth. The gain is assumed very large and the output is a voltage generator of zero impedance providing an EMF of A_v times the difference between the input levels. Clearly, when $V_1 = V_2$ the output will be zero. Opamps are usually depicted in the forms shown in *Figure 5.1(b)* and *(c)*, which show the differential and single-input modes respectively.

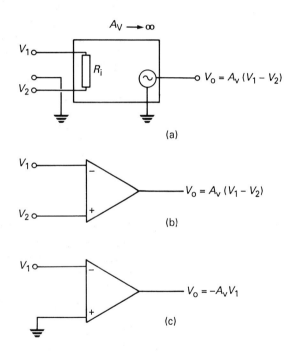

Figure 5.1

These inputs are marked + and −; the positively marked terminal is known as the non-inverting input, and the negatively marked terminal is known as the inverting input. These terms are self-explanatory. A signal applied to the inverting terminal, with the other input earthed, comes out antiphase to the input; that is, there is an effective 180° phase shift through the amplifier. A signal applied to the non-inverting input comes out in phase with the input. When feedback is applied, it must always be connected to the inverting input for negative feedback, but the signal may be applied to either terminal depending upon the application.

THE DIFFERENTIAL AMPLIFIER

The input circuit of an opamp is a sophisticated version of the so-called emitter-coupled amplifier or differential amplifier, shown in its basic form in *Figure 5.2*. Two identical transistors have their emitters connected together and returned to the common rail by way of resistor R_E. For clarity, biasing arrangements are omitted. The name 'differential amplifier' stems from the fact that the input signal v_i is balanced with respect to earth so that the two base inputs are antiphase to each other. As the forward bias of transistor T_1 is increased, the forward bias of transistor T_2 is reduced, and conversely. If the transistors are perfectly matched, the current flowing in R_E will remain constant. Hence, although this resistor is not bypassed, there is no negative feedback voltage derived across it as there is in the single common-emitter amplifier.

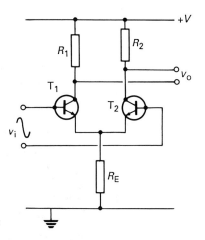

Figure 5.2

Like the input signal, the output signal is also balanced with respect to earth. Since the collector voltages swing in opposite directions, the output voltage taken between the collectors for antiphase inputs is double that of each transistor. This is known as the *differential mode* of operation.

Suppose now that in-phase signals are applied to the two input terminals. The collector voltages swing in the same direction and the output signal is, assuming a perfectly balanced circuit, the zero difference between the two collector potentials. This is known as the *common mode* of operation.

We can now apply these input and output conditions of the differential amplifier to the operational amplifier. Here we find we have the two input terminals but only one output terminal. This is because a double-ended to single-ended conversion stage is included in the opamp. There is also a final output amplifier stage. *Figure 5.3* shows a simplified operational amplifier where the various stages are easily recognized: the differential input stage A, the conversion stage B, and the final amplifier C. The differential input stage uses Darlington pairs, and the emitter R_E would be replaced in practice by a constant-current transistor. The balanced to unbalanced converter is a *p-n-p* transistor with its base–emitter junction connected directly across the output

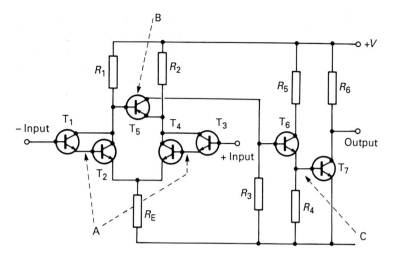

Figure 5.3

collectors of the input amplifier; its collector current is then determined by the potential difference between these points. A more or less conventional output amplifier then follows. Although much more sophisticated in design, most opamps are simply versions of this arrangement.

We can now sum up in a form illustrated in *Figure 5.4*. At (a), two in-phase inputs are applied and the output is the amplifier gain A_v times the difference between the inputs. If $v_1 = v_2$, then the output is zero. In (b) two antiphase inputs are applied and the output is the amplifier gain A_v times the sum of the inputs. For single-ended input conditions, we turn to diagrams (c) and (d). At (c) an input is applied to the inverting input; the output is an inversion of the input and is given by $-A_v v_1$. At (d) an input is applied to the non-inverting input; the output is in phase with the input and of amplitude $+A_v v_1$.

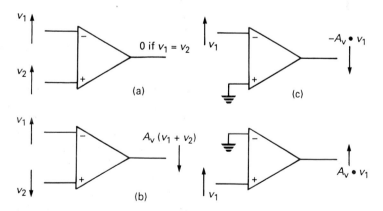

Figure 5.4

Ideally, the output voltage of an opamp should be zero when the input voltage is zero. This would be true if everything was perfectly matched inside the amplifier – a situation we have assumed in the previous analysis. In real amplifiers such matching is impossible to achieve (although circuits fabricated on a single silicon chip can come very close to the ideal), and the output is not a precise zero when the input is zero. An offset voltage can be applied to opamps to compensate for this error and to set the output to zero with the inputs earthed.

The opamp should also ideally ignore any common-mode signals; that is, if an identical signal is applied to both the inverting and the non-inverting inputs there should be no change in the output. In practice there is always some variation, however small, and the ratio of the change in output voltage to the change in common-mode signals is called the common-mode rejection ratio (CMRR). It can be proved that the best CMRR rejection is obtained when the common-emitter resistor of *Figure 5.2* is large in value. This makes design difficult, but the problem is overcome in practice by replacing this resistor with a constant-current source. In this way the *effect* of a high impedance is achieved with a very small voltage drop.

It is usual to power opamps by both a positive and a negative supply with the circuit earth line at the centre of the supply. In this way the output is enabled to swing both above and below ground, although this facility is not always required. Most opamps are available in an eight-pin dual-in-line integrated package and the connections are usually as given in *Figure 5.5(a)*. The power supply is connected between pin 7 (positive) and pin 4 (negative), although the latter pin may go to earth for unbalanced feeding. The output is at pin 6, and pin 2 is for the inverting input and pin 3 for the non-inverting input. In addition there are two 'offset null' inputs at pins 1 and 5, which are used (when necessary) to bring the output to zero when the inputs are zero. The way in which the various inputs are applied is shown in *Figure 5.5(b)*. The maximum permissible supply voltage for a particular opamp will be found in the manufacturer's recommendations.

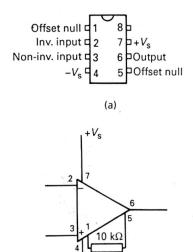

(a)

(b)

Figure 5.5

BASIC CONFIGURATIONS

All opamps have a very high voltage gain (up to 100 dB) in the open-loop configuration – that is, if the gain is measured between input and output without the addition of any kind of feedback. As it stands this kind of gain is of no value, because the smallest potential difference of a fraction of a microvolt between the input pins will send the output into saturation and the output will swing as far as it can towards one or other of the supply levels. It may be asked why such an enormous gain characteristic is provided, since quite clearly it cannot be used as such. The point is that heavy negative feedback can be introduced, with all its advantages, and still leave us with as much useful gain as we require.

The inverting amplifier

The circuit arrangement of *Figure 5.6* belongs to the general category of inverting circuits. The common feature of these circuits is that the non-inverting input is connected to earth. Both the input signal and the feedback signal are applied to the inverting input. The most important thing about this circuit is that whenever some of the output is fed back to the inverting

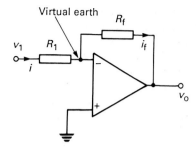

Virtual earth

Figure 5.6

input (negative feedback) the voltages measured on both the inverting and the non-inverting inputs will be the *same*. The output of the opamp will adjust itself, either producing a current or absorbing a current to keep these voltages the same.

When the non-inverting input is earthed, therefore, the inverting terminal will be held at 0 V. Since this input is at 0 V but not actually connected to earth, it is referred to as a *virtual* earth. Because of this, the full input voltage v_1 will develop across the input resistor R_1, and the signal current in R_1 will be $i_1 = v_1/R_1$. Now consider the feedback resistor R_f which is connected between the virtual earth and the output; the current in R_f will be $i_f = v_o/R_f$. Since negligible current will flow into the opamp because of its high input resistance, we may think of the input circuit as a source of current which must flow into R_f. Hence $i_1 = -i_f$, so

$$\frac{v_1}{R_1} = \frac{-v_o}{R_f}$$

and the closed-loop gain as given by

$$A_v' = \frac{v_o}{v_i} = \frac{-R_f}{R_1}$$

Note that the gain includes the sign inversion and that its magnitude is determined *solely* by the ratio of the external resistors. This expression for gain is strictly only true if the open-loop gain is infinite; for an opamp in which gains of 10^5 are usual, the expression is very closely accurate for most practical purposes.

> 1 What voltage would you measure at the inverting input of an opamp with negative feedback if the non-inverting input was taken to -5 V?
> 2 If, in Figure 5.6, $R_1 = 10$ kΩ and $R_f = 100$ kΩ, what would be the closed-loop gain of the amplifier?

The summing amplifier

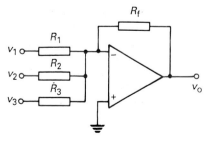

Figure 5.7

If the basic inverting circuit is modified to accept additional signal inputs and 'summing' resistors are added, the amplifier performs the mathematical operation of addition. *Figure 5.7* shows such a circuit system and, because of its ability to add, the virtual earth point is often referred to as the *summing* point. Let the input voltages be v_1, v_2 and v_3. Then the currents in the input resistors are

$$i_1 = \frac{v_1}{R_1} \qquad i_2 = \frac{v_2}{R_2} \qquad i_3 = \frac{v_3}{R_3}$$

All these input signal currents flow into R_f, generating an output voltage given by

$$v_0 = -(i_1 + i_2 + i_3)R_f$$

$$= -\left(v_1\frac{R_f}{R_1} + v_2\frac{R_f}{R_2} + v_3\frac{R_f}{R_3}\right)$$

If the three resistors are equal in value and equal also to R_f, the voltages v_1, v_2 and v_3 will be added together at the output. For unequal values, the output will depend upon the ratios of the various input resistors to the feedback resistor, and v_o will be a 'weighted' average of the input. The circuit therefore behaves as an *analogue* adder; we will be returning to this circuit when we investigate analogue-to-digital converters later on.

The non-inverting amplifier

The inverting circuit just discussed can be realized with either single-ended or differential-input amplifiers. Non-inverting amplifiers in general require a differential input. The basic non-inverting amplifier is shown in *Figure 5.8*. Input is applied to the non-inverting terminal and feedback from the output is applied to the inverting terminal. As shown, the whole of the output is fed back, but this is a specialized case and in general a fraction of the output is used. To find the gain of this circuit we note that

$$v_o = A_v(v_1 - v_o)$$

so that

$$v_o = \frac{v_1}{1 + 1/A_v}$$

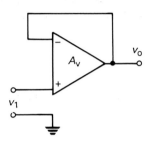

Figure 5.8

Since A_v is very large we can take $1/A_v$ to be zero; then $v_o = v_1$ and the circuit gain is unity.

The output voltage will therefore always take on the value required to drive the signal between the inverting and the non-inverting inputs towards zero. This is a voltage follower, a more sophisticated version of the emitter or source follower.

The voltage follower is, of course, a special and very useful case of the non-inverting amplifier, as already indicated. In the general form, shown in *Figure 5.9*, only a fraction of the output is fed back to the inverting terminal. From simple feedback theory we have

$$\beta = \frac{R_1}{R_1 + R_f}$$

and for A_v very large,

$$A_v' \rightarrow \frac{1}{\beta} \rightarrow 1 + \frac{R_f}{R_1}$$

The overall gain is now greater than unity, is positive and is determined solely by the values of R_1 and R_f.

It is useful to remember that in the non-inverting circuit the closed-loop gain is given very closely by $1/\beta$ and in the inverting circuit by $(1 - 1/\beta)$.

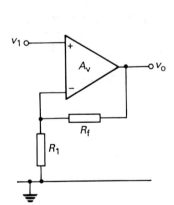

Figure 5.9

---

3 An opamp has a closed-loop gain of 100. Calculate the required values for R_f for (a) an inverting configuration (b) a non-inverting configuration, given that the input resistor (R_1) is 47 kΩ.

The differential amplifier

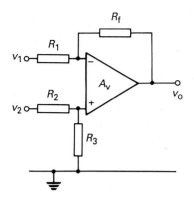

Figure 5.10

The summing point principle applies as much to the differential mode of operation as it does to the single-ended input amplifier. The circuit is shown in *Figure 5.10*. This configuration is widely used in instrumentation, and has the advantage that in-phase or common-mode signals at the inputs produce no variation at the output. Such in-phase signals include thermally generated noise, DC fluctuations and drift. The differential amplifier therefore discriminates against these unwanted intruders and responds only to significant (wanted) signal variations.

If we assume an infinite open-loop gain and infinite input resistance then the inverting and non-inverting input points will be at the same common-mode input voltage. Hence, as we have already noted,

$$v_o = -\frac{R_f}{R_1}(v_2 - v_1)$$

For optimum performance, the resistances to earth should be matched with

$$\frac{R_f R_1}{R_f + R_1} = \frac{R_2 R_3}{R_2 + R_3}$$

It is of some importance to note that there is a marked difference between the input resistances of the two input points. At the inverting input we see the virtual earth point; hence the input resistance is simply R_1. At the non-inverting input we see R_2 in series with R_3 since no current flows into the non-inverting terminal itself. Where high gain is concerned, so that R_f is considerably greater than R_1 (and the same ratio applies to R_3 and R_2), the discrepancy between the input resistances can be very large. This can lead to excessive noise pickup on the high-resistance input point (the non-inverting input) relative to that on the low-resistance input point. One way of overcoming this problem is to precede the differential amplifier with two unity-gain voltage followers as seen in *Figure 5.11*. This maintains a very high input resistance on both inputs.

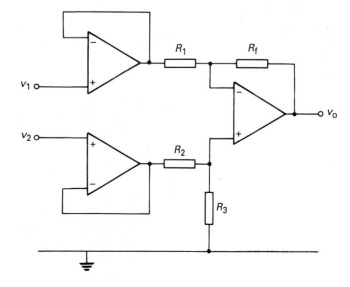

Figure 5.11

Example 1
In the circuit of Figure 5.10, $R_f = R_3 = 0.5$ MΩ,
$R_1 = R_2 = 10$ kΩ. What is the amplifier gain, and what is
the resistance seen at each input?

The gain $A_v' = R_f/R_1 = 50$.

At the v_1 input point we see only the effect of R_1; hence
the input resistance here is 10 kΩ. At the v_2 input point we
see the effect of R_2 and R_3 in series, hence the input
resistance here is 510 kΩ.

Example 2
In the operational summer shown in Figure 5.7, suppose
the input voltages to be $v_1 = 2$ V, $v_2 = 3$ V, $v_3 = 1$ V.
also suppose $R_1 = R_f = 10$ kΩ, $R_2 = 15$ kΩ, $R_3 = 20$ kΩ.
What will be the output voltage?

Using the expression we derived for the output of the
operational summer and substituting values, we have

$$v_o = -10 \left(\frac{2}{10} + \frac{3}{15} + \frac{1}{20} \right)$$

$$= -4.5 \text{ V}$$

RATED OUTPUT

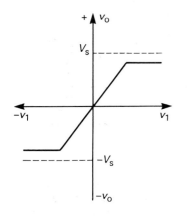

Figure 5.12

There is clearly a limitation to the output voltage swing
obtainable from an opamp. These limitations are specified by
stating the rated output voltage v_o, and current i_o, for which
linear operation applies. The limit on the output voltage is the
saturation voltage, which is normally a volt or so in excess of
the rated output voltage. Although you won't damage the
opamp by driving it into saturation (as indeed it often is,
particularly in non-linear and switching applications), the
important thing is that the recovery time may be long. This can
affect the speed of switching.

An experimental procedure is given later on which you can
plot a typical transfer characteristic relating the input to the
output voltages of an opamp. The sort of graph you might
expect to emerge is shown in *Figure 5.12*. Note that the gradient
of this characteristic is equal to the amplifier closed-loop gain.
Since the amplifier operates on finite values of power supply,
typically between 9 V and 15 V, the characteristic exhibits a
saturation limit slightly below the supply level.

FREQUENCY CONSIDERATIONS

An operational amplifier is a multistage circuit having a dozen
or more integrated transistors in its makeup; it has also a very
high open-loop gain. In use, of course, negative feedback is
applied and the gain is reduced to the manageable levels
required. But this does not guarantee that the circuit will be
stable. At high frequencies, the open-loop gain falls because of
the internal capacities, and these also introduce internal phase

shifts which modify the phase relationships of 0° or 180° assumed for the non-inverting and inverting configurations at low frequencies. In many cases, some sort of frequency compensation has to be applied to the circuit system to maintain stability over the desired operating range. The manufacturer's sheets must be consulted for this, but usually a small-value capacitor (10−47 pF) connected between pins 1 and 8 on a number of available opamps takes care of things for frequencies up to some 200 kHz, irrespective of the amount of feedback.

Stability in the inverting configuration, when gains greater than 1000 are contemplated, can be improved by shunting the feedback resistor R_f with a capacitor equal in value to $R_1 C_i / R_f$, where C_i is the input capacitance of the opamp at the inverting terminal.

AN EXPERIMENT

In this experiment you will be able to see the effect of compensating for offset as well as to plot a transfer characteristic curve of output against input voltage. Set up the circuit shown in *Figure 5.13*, using a 741 or 741S opamp. The supply can be anything between ±9 V and ±15 V and may be obtained from batteries (good ones!) or a balanced stabilized supply. Make sure you connect the mid-point (zero line) of the supply to the common (earth) rail.

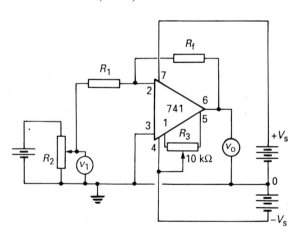

Figure 5.13

Connect the input terminals (pins 2 and 3) together and to earth with a short piece of wire so that the input voltage v_1 is zero. Adjust the offset null potentiometer R_3 until the output voltage v_o (preferably measured on a digital voltmeter) is also zero. Hold your finger on the opamp for a while; you will probably find that the output shifts slightly from the zero condition. This shows that the offset is temperature sensitive, so try to avoid any such temperature variations during the rest of the experiment.

Make R_f = 100 kΩ, R_1 = 10 kΩ, so that the gain is −10; use 1 per cent resistors if you can get them. Remove the short-circuit you put on the input and apply an increasing range of voltage inputs by adjustment of potentiometer R_2, noting the output voltage v_o at each step. Now reverse the input battery, and repeat the above using a series of negative input voltage steps. From your tabulated results, plot a graph of the *measured* values of v_o against the *calculated* values, i.e. $v_1 \times 10$. Your graph should be of the form shown in *Figure 5.12* but with the horizontal axis replaced by the calculated values of gain. Is the curve completely symmetrical? If not, can you suggest a reason why not? How does the saturation level relate to the applied supply voltage?

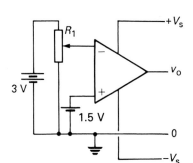

Figure 5.14

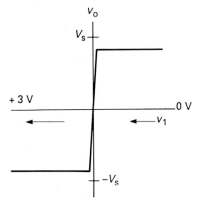

Figure 5.15

Example 3
An opamp is connected up as shown in *Figure 5.14*. The non-inverting input is held at a steady DC level by a 1.5 V battery supply (which we might call the reference voltage), while the inverting input can be adjusted between 0 and 3 V by potentiometer R_1. Explain what happens to the output voltage v_o as R_1 is adjusted from zero to 3 V.

As the inverting input is increased from zero towards the mid-point of the potentiometer supply ($=1.5$ V), v_1 is less than the reference level; hence the output will be positive. Since the amplifier is operating at its very large open-loop gain value, any voltage difference at the inputs will drive it into saturation; hence the output will be at its maximum positive saturation level. As soon as the slider of R_1 passes the mid-point, v_1 will be greater than the reference level; hence the output will switch immediately to its maximum negative saturation level. A transfer characteristic for this circuit is shown in *Figure 5.15*.

An opamp used in this way is known as a zero-crossing comparator. It compares one input with a reference level and switches the polarity of its output when the input voltage passes through the reference level.

A simple square-wave generator can be made by applying a small sine-wave input to the v_1 position, the reference level simply being earth.

4 List the ideal properties of an operational amplifier.
5 How can an operational amplifier be used as an analogue adder?
6 A voltage follower has unity gain and is non-inverting. What advantages does such an amplifier have? Explain briefly how it works.
7 What is a differential amplifier? Name its advantages.
8 Distinguish between a differential and a common mode of operation. What is meant by the common-mode rejection ratio?
9 Three voltages, v_1, v_2 and v_3 are to be added by using the operational amplifier circuit of Figure 5.7. If each resistor is 100 kΩ, what must be the value of R_f such that $v_o = v_1 + v_2 + v_3$?

10 What will be the output voltage if, in the circuit of Figure 5.7, $v_1 = 2$ V, $v_2 = 3$ V, $v_3 = 4$ V, $R_1 = 10$ kΩ, $R_2 = 22$ kΩ, $R_3 = 12$ kΩ and $R_f = 47$ kΩ?

11 How might an operational amplifier be used to compare two voltage levels and distinguish which is the greater?

6 DA and AD converters

Aims: At the end of this unit section you should be able to:
Explain methods of converting analogue signals to digital signals, and vice versa.
Draw circuits and analyse the operation of R-2R ladder-type digital-to-analogue (DA) converters.
Be familiar with the basic characteristics of converters. Explain the operating principles of analogue-to-digital (AD) converters.

The world we live in is an analogue world; things operate 'smoothly' and 'continuously' in the sense that there are no sudden jumps in the natural course of events. When changes occur in physical happenings, even though they may sometimes seem abrupt, there is a continuous transition of state throughout the time the changes are taking place. Time itself is continuous; we can divide it up into minutes, seconds, even microseconds, but it is there 'all the time', filling in the gaps between any two given instants, however close those instants are together. So we say time is an analogue quantity, just as things like velocity, pressure and temperature are analogue. Analogue quantities, in other words, can take up *any* value within a range of values. An analogue meter, for example, has a pointer which can move to *any* position along its scale length; whether we can actually read it to a given degree of accuracy is immaterial. The pointer can take up any one of an infinity of positions.

However, quite a proportion of the electronics scene has now been taken over by digital devices: computers, video and compact disc recordings, control systems and so on. When we want to store the output from an analogue device, such as say a varying voltage or current, in a computer memory, the data must be converted from its analogue (continuous) form to a digital (discontinuous) form of 0s and 1s which the computer will accept in a suitably coded way. To facilitate the transition, an analogue-to-digital converter (ADC) is required. An ADC produces a pulsed output in which each input voltage or current variation is converted into a corresponding binary word, this taking place probably as often as 10^4 times per second.

Conversely, whenever a digital signal needs to be represented in analogue form, a digital-to-analogue converter (DAC) must be used. This device produces an output voltage or current proportional to the magnitude of the binary number applied to the input. *Figure 6.1* gives some idea of this sort of conversion, though keep in mind that the vertical scale represents a discontinuous quantity. As DACs are the easier to understand, we begin with an examination of the fundamental principles behind these, leaving the rather more complicated ADC till later on.

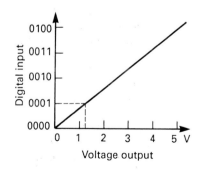

Figure 6.1

BINARY-WEIGHTED DA CONVERTER

There are a number of digital-to-analogue converters, and in some simple applications a type based on the summing mode of an operational amplifier is adequte. We have already covered this subject in Section 5, but a recapitulation at this point will be useful. *Figure 6.2* shows what is known as a binary-weighted resistor ADC. We have seen earlier how currents flowing through resistors $R_1 - R_4$ are summed at the virtual earth point (the inverting input terminal) and converted to a proportional voltage by the opamp and the feedback resistor R_f.

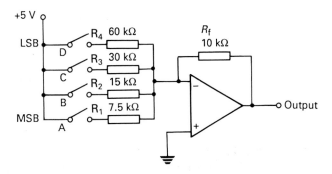

Figure 6.2

If a four-bit binary word (or number) A, B, C, D is applied to the input resistors by way of the matching switches, the resistors produce binary currents 'weighted' according to the value of the word. This is accomplished by giving each resistor in the array a value *inversely* proportional to the significance of the input bit of the applied digital word. Thus the resistor connected to the most significant bit (MSB) may have a value R; the next resistor will then be $2R$, the next $4R$, and so on.

For example, suppose the input word is the binary number 0101, equivalent to decimal 5. Then switches B and D will be closed, switches A and C open. Notice that the information is entering in parallel form. With the values of resistors indicated, the current through R_2 will be $5/(15 \times 10^3)$ A = 334 μA, and through R_4 will be $5/(60 \times 10^3)$ A = 83.3 μA. The total current flowing into the summing point and hence through R_f will be the sum of these two, or 417.3 μA. The output voltage, therefore, will be $10^4 \times 417.3 \times 10^{-6}$ = 4.17 V. (Strictly we should append a negative sign to this output, but the argument will be unaffected by its omission.) Now suppose the binary input to change to 1000 or decimal 8. Then only switch A will be closed; the input current to the summing point will be $5/(7.5 \times 10^3)$ A = 667 μA, and the output voltage will change to $10^4 \times 667 \times 10^{-6}$ = 6.67 V.

Try a few more combinations for yourself. You will find that the output increases in fifteen steps, each step of amplitude 0.833 V, as the binary input ranges from 0000 to 1111 (decimal 15). By feeding the inputs from a four-bit binary counter and displaying the output on an oscilloscope, a staircase waveform will be obtained showing each step in the progression, rather as in *Figure 6.3*. What we have, therefore, is a voltage output (the analogue signal) proportional to the (digital) input magnitude. Notice that the opamp performs as a current-to-voltage converter.

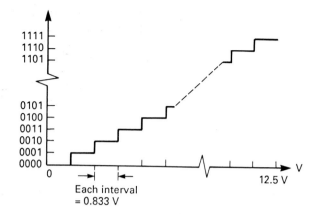

Figure 6.3

R-2R LADDER DA CONVERTER

The binary-weighted converter has the disadvantage that a separate resistor is required for each input bit. Anything over some six bits leads to a range of resistance which is too great for proper operation. A twelve-bit device, for example, assuming that we begin with $R_1 = 10$ kΩ corresponding to the MSB, would require a 20.48 MΩ resistor for the LSB. Values such as this cannot be manufactured with the degree of accuracy and stability which is necessary.

An alternative system, which produces binary-weighted currents but which uses only two values of resistance, is shown in *Figure 6.4*. As in the previous case, the weighted currents are converted to proportional voltages by the opamp and feedback resistor R_f. The 5 V reference line is illustrative only; commercially available converters using the *R-2R* ladder operate with other levels. The switching is, of course, electronically controlled and operated (in this example) by the four-bit parallel binary input.

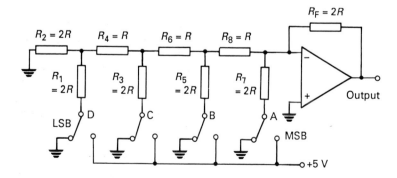

Figure 6.4

In theory the actual values chosen for the resistors are unimportant, provided that all the *R*s have the same value and the 2*R*s are double this. In practice the values selected should be low enough to provide a low output impedance but not so low as to put appreciable loading on the reference line. Typical values are $R = 10$ kΩ, $2R = 20$ kΩ. To follow the operation, suppose that switch A (representing the MSB) is connected to the reference line and the other switches to earth. Then R_1 (=2R) and R_2 (=2R) are in parallel to earth; their equivalent resistance is then simply equal to R. This equivalent R adds to R_4 (=R) to form another $2R$ in parallel with R_3 (=2R) and earth (see *Figure 6.5(a)*). The combination of R_3 (=2R) in parallel with the previous two equivalent Rs (in series) reduces to a single R in series with R_6 (see *Figure 6.5(b)*). Carrying on in this way (do it for yourself) leads to the circuit shown in *Figure 6.5(c)*.

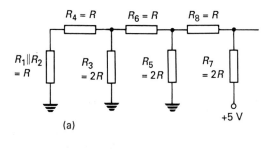

(a)

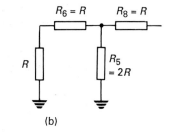

(b)

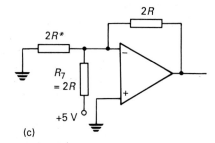

(c)

Figure 6.5

When you have recovered from this traumatic analysis, a study of Figure 6.5(c) reveals that no current will flow through the final equivalent resistor (designated as $2R^*$) because one end is connected to earth and the other end to the virtual earth of the opamp. This resistor therefore plays no part in the further analysis. The other resistor, R_7, connects to the 5 V reference line; hence, taking its value as 20 kΩ, a current of $5/(20 \times 10^3)$ A or 0.25 mA flows through R_7 and the feedback resistor R_f, which is also equal to $2R$ or 20 kΩ. The output voltage produced is therefore 5 V, representing the analogue of the MSB on the binary input.

It is left as a self-exercise for you to analyse the circuit when switch B is closed to the reference line and A, C and D are earthed. This represents the next most significant digit at the input. You should end up with a circuit equivalent to that shown in *Figure 6.6*, where the current into the summing point is $2.5/(20 \times 10^3)$ A or 0.125 mA. The output voltage is then 2.5 V.

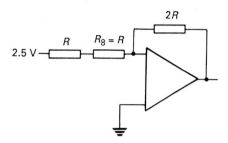

Figure 6.6

Example 1
If all four bit switches are closed in Figure 6.4, what will be the output voltage?

Considering the above analysis, the analogue outputs for the four inputs (commencing with the MSB) are clearly 5 V, 2.5 V, 1.25 V and 0.625 V. With all switches closed, the sum of these will be 9.375 V.

CHARACTERISTICS OF DA CONVERTERS

There are a number of characteristics associated with DACs which need to be known if full use is to be made of application notes issued by the manufacturers.

The first of these is the *resolution* of a converter. This is a function of the number of bits in the input data. A four-bit converter, for instance, has 2^4 or 16 output (or quantization) levels, so its resolution is said to be 1 part in 16 or 6.25 per cent.

The accuracy (or *quantization error*) of a converter is defined in terms of the quantization intervals or the number of steps to be found in the staircase output. The number of steps for an N-bit converter is clearly (2^N-1); hence the analogue output can be no more accurate than one part in this number. If the total output excursion of the converter is known, then the quantization error is expressed as plus or minus half the quantization interval. So, referring to Figure 6.3 as an example, where the total voltage range is 12.5 V, the converter has an error of $12.5/(2 \times 15) = \pm0.41$ V, there being fifteen intervals for a four-bit device.

An eight-bit converter will have 255 intervals, and for the same voltage range the error would be $12.5/(2 \times 255)$ or ±0.245 V. So the greater the number of bits, the smaller the quantization error.

The *linearity error*, which might easily be confused with quantization error, is a measure of the amount by which the actual output differs from a mean straight line (a) drawn along the staircase as shown in *Figure 6.7*. A linearity error curve is illustrated by the broken line (b), while an error resulting from a scaling (or gain) inaccuracy is shown in line (c). This problem can often be traced to the feedback resistor across the opamp being incorrectly selected. Finally, line (d) which runs parallel to the wanted output line (a) comes about because of offset error, the output not being zero when all the inputs are zero. Careful offset adjustment in the opamp will take care of this problem in most cases.

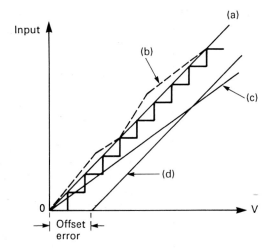

Figure 6.7

Output *settling time* is another parameter you will find on application sheets. This is defined as the time the converter needs to settle within ±0.5 per cent LSB of the final value after the input data changes. It is usually specified with all bits switched on or off at room temperature. A typical figure is 0.1 μs. The converter is limited, therefore, in the handling of inputs whose period is less than the settling time.

BIPOLAR AND UNIPOLAR DA CONVERTERS

So far we have discussed DACs which have provided a single-polarity output. Such converters are termed *unipolar*. Converters are often required which give an output running across both positive and negative values, and these are known as *bipolar* converters. It is usual on these systems to have both positive and negative reference lines instead of the positive reference and earth so far discussed. A typical bipolar staircase output is shown in *Figure 6.8*, and this should be compared with Figure 6.3.

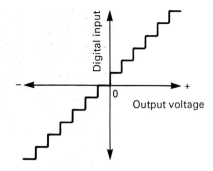

Figure 6.8

PRACTICAL EXPERIMENTS WITH DA CONVERTERS

Practical DACs, available in integrated circuit form, often have reference levels which are built in, and the values are chosen such that a direct correlation exists between the output voltage and the decimal equivalent of the binary input. An eight-bit system having a reference of 2.55 for example, which consequently has an input range of 0 to 11111111 (or 0 to 255 decimal) will have an output of 0.01 V per digit input. An input of 187 (decimal equivalent) will therefore produce an output of 18.7 V.

An integrated DAC suitable for experiments is the ZN428, a systems diagram of which is given in *Figure 6.9*. This is a typical 8-bit monolithic converter which, in addition to the *R-2R* ladder network and an optional internal 2.5 V reference, has input latches which will hold the input data when enabled. ('Monolithic' means that all the system is on one single chip.) This facilitates updating from a computer bus line, for instance. The input is TTL and 5 V CMOS compatible. The switching system is transistor operated.

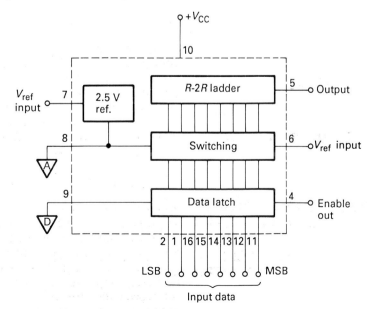

Figure 6.9

A practical circuit design is given in *Figure 6.10*, and from this a number of experiments can be derived. The eight-bit input can be driven from TTL logic outputs. The included reference source requires an external load resistor R_1 and a decoupling capacitor C_1. In addition, an opamp (for which the ubiquitous 531 is suitable) is needed to provide both buffering and gain. An offset-null trimmer control is also necessary to get optimum accuracy at low-output voltage levels, and the potentiometer R_3 is adjusted for the required full-scale output with all inputs held high. The enable input should normally be set low to give a transparent latch. The system provides a unipolar output and, if the given voltage levels are used, the output will have a full-scale range of +5 V.

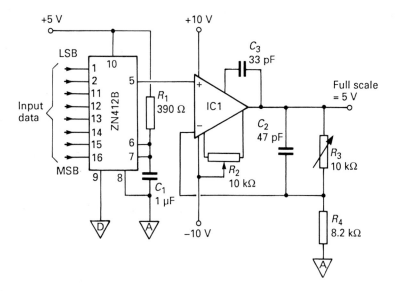

Figure 6.10

Setting up is quite easy. Put all input bits to 0 (low) and adjust the offset control to give zero output. Then put all input bits to 1 (high) and adjust R_3 to give an output of 4.98 V, that is, full scale less 1 LSB.

ANALOGUE TO DIGITAL

The ADC is the inverse of the DAC already discussed. DACs are much simpler than ADCs, but nevertheless a number of types of ADC use the techniques of DAC systems as parts of their circuitry. We will examine three common types of ADC; most practical circuits depend upon various arrangements and combinations of these.

PARALLEL AD CONVERTER

This form of ADC, also known as a simultaneous or flash converter, has not only the simplest circuitry but also the fastest operating speed. We can explain the operation of the parallel converter by looking at *Figure 6.11*. Three opamps are used as comparators; threshold voltages are applied to the inverting inputs through a potential divider chain made up of resistors R_1

to R_4. The input or reference (V_{ref}) voltage to this chain (2 V in our example) represents the full scale or maximum of input level. In operation, each comparator compares the analogue input voltage with its own particular reference voltage. When the divider resistors are made of equal value, the voltages at each junction (the inverting inputs) increase by one-quarter of the total reference voltage, that is there are increments of 0.5 V from 0 to 2 V. The input analogue signal to be digitized is then applied to all three non-inverting terminals in parallel.

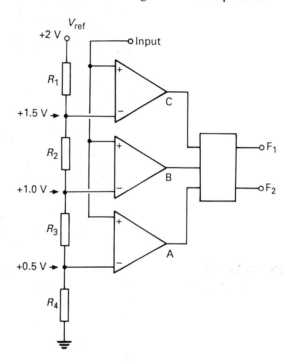

Figure 6.11

The output of each comparator will therefore go high if the input voltage on its positive input is greater than the reference level on its negative input. Thus, for example, for an input voltage of 1.3 V and the reference of 2 V, comparators A and B will have high outputs (logic 1) and comparator C will have a low output (logic 0). An input less than 0.5 V will not trip any of the comparators; an input greater than 1.5 V will trip all three comparators. As this output code is not binary, it is necessary to have an encoder following the comparators so that the various output combinations of high and low can be converted into binary. This encoder may be made up from a number of logical gates (something you might try for yourself), or a programmed ROM can be used.

A truth table showing the voltage input range, the comparator outputs and the digital outputs is given in *Table 6.1*.

A converter such as this, having only three comparators, will resolve an input voltage to one only of four input levels. As the table shows, this is equivalent to two binary bits of resolution.

Table 6.1

Input (V)	Comparator output A	B	C	Binary F_1	F_2
0–0.5	0	0	0	0	0
0.5–1.0	1	0	0	0	1
1.0–1.5	1	1	0	1	0
1.5–2.0	1	1	1	1	1

To extend the coverage, a greater number of reference levels, and comparators, are needed. Each reference voltage corresponds to a quantization interval, so for three binary bits of resolution there must be seven reference levels and hence seven comparators. For N-bits of resolution, the number of comparators is (2^N-1), so an eight-bit converter requires 255 comparators and references. This is the only major drawback to the parallel ADC.

The accuracy is completely dependent upon the reference levels and hence upon the accuracy of the resistors used in the divider chain. The operating speed is very high and these converters are therefore used for the digitization of analogue voltages from high-frequency sources; hence the name 'flash' converter. A typical eight-bit converter can produce a digitized output within 50 ns of the input being applied.

COUNTER-TYPE AD CONVERTERS

Possibly one of the simplest types of ADC, in both economy of components and understanding, is the single-counter converter which includes in its makeup a DAC of the kind already covered. *Figure 6.12* shows a block diagram of the system. In addition to the DAC, there is a binary counter and a comparator. There may also, in practical circuits, be a latching system so that the binary output can be retained in the latch until required.

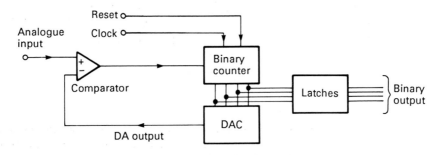

Figure 6.12

At the start of the conversion sequence we assume that the analogue input is 0, that a reset signal has been applied to the counter and that the output of the DAC is 0. The inverting input to the comparator is therefore also at 0. When some voltage is applied to the non-inverting terminal, the comparator output goes high and enables the counter, which begins to count the clock pulses. Each clock pulse advances the counter one step and the input to the DAC consequently increases. The output from the DAC therefore increases also, by one step for each clock pulse. This output feeds into the comparator and, as long as this voltage is below the analogue input, the output from the comparator will remain high. As soon as the DAC output exceeds the analogue input level, however, the comparator output will trip low and shut off any further clock pulses to the counter. The digital output will then be held in the latches at a value corresponding to the analogue voltage level at the input. The counter is then reset and the sequence begins again.

The single-counter type of converter is a relatively slow operator, being dependent upon the speed of the counter and the DAC. A very precise DAC is also essential. Since the count has to start at 0 for each conversion cycle, the time for each conversion depends upon the amplitude of the analogue signal. Further, during a conversion, the DAC output may correspond exactly to, or be very slightly below, the analogue level at the comparator; the comparator will consequently fail to trip until the counter has moved through a further clock cycle. The greatest quantization error is therefore one complete interval. This can be reduced by the application of an offset current to the output of the DAC.

An improvement on the single-counter converter is the *tracking* or *follower* converter in which the up-only counter is replaced by an up-down counter. This counts up when the output from the comparator is high, and counts down when the output from the comparator is low. The clock pulses are routed to the count-up or count-down inputs by the gate switches on the comparator output. *Figure 6.13* shows a block diagram of this system.

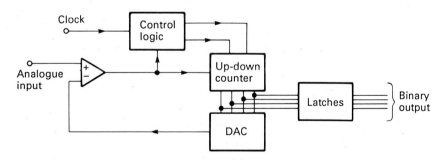

Figure 6.13

At the start of a conversion, assume that the DAC output is at 0. Then with some analogue input voltage, the output of the comparator will be high. This routes the clock pulses to the count-up input of the counter, and the output of the DAC increases until it exceeds the analogue level. The comparator then trips low, the clock is transferred to the count-down input of the counter, and the counter begins to count down from its previous figure until the correct level is reached. If the input has remained unchanged, however, the first one-down count will drop the output from the DAC immediately below the analogue level, so tripping the comparator output high and initiating the reversal of the count. The following one-up count will again be sufficient to trip the comparator, and the one-up, one-down sequence will continue for as long as the analogue input remains steady. This is a disadvantage of this type of converter; we get an oscillatory least-significant-bit output for an unvarying input.

SUCCESSIVE APPROXIMATION

The successive approximation (SA) type of converter has the advantage over the previous systems in that N bits of resolution can be secured with only N clock pulses. In essence, the method follows on from the counter-type converters except that the counter is replaced by a register and some addition control circuitry is included between this register and the comparator. The ZN427E is a commercially available eight-bit converter of this sort. A simplified block diagram of an SA converter is shown in *Figure 6.14*.

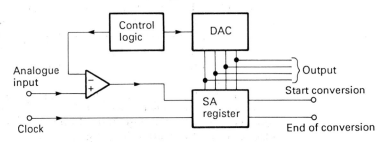

Figure 6.14

The system operates as follows. A reset signal (the start-conversion pulse) sets all bits in the register to 0. The DAC output is thus 0, and any input signal at the analogue terminal causes the comparator output to go high. On the application of the first clock pulse, the register turns on its MSB to the DAC; all other bits remain at 0. This produces a voltage from the converter (V_1). This voltage is compared with the input voltage, and a decision is made on the next clock pulse edge to set the MSB to 0 if V_1 is greater than the input or retain a 1 if V_1 is less than the input. In either case, on the next clock edge the register will set the next MSB. In turn it will keep or reset this second bit, again by comparing the DAC output with the input. This process is repeated until the LSB is reached. A bit is retained if V_1 is less than the input and a bit is reset if V_1 is greater than the input. When all the bits have been tried, the register end-conversion output goes high; the digital output from the converter is then a valid representation of the input voltage. This output is latched until the next start-conversion pulse.

If this description has been difficult to visualize, you might compare the successive approximation method with a way of finding the mass of something by using weights of 64, 32, 16, 8, 4, 2 and 1 kg. The most significant weight is tried first, and if this is too small a 1 is recorded. Then the 32 kg weight is tried by adding it to the 64 kg weight, and so on. If any added weight makes the total *greater* than the object being weighed, the weight is put aside and a 0 is recorded. If any added weight makes the total *less* than the object being weighed, a 1 is recorded. Each weight needs only one trial. When they have all been tried and a balance obtained, we will have a binary word which represents the weight of the object.

In the electrical converter, the total conversion time is equal to N clock pulses, where N is the number of bits in the register. This time does not depend upon the amplitude of the analogue input as it did with the previous circuits. *Figure 6.15* shows a conversion cycle for an eight-bit successive approximation converter. A timing diagram for such a converter with the digital word 01101010 as output is shown in *Figure 6.16*.

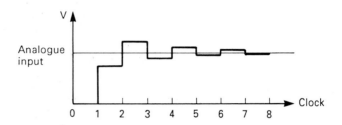

Figure 6.15

The twelve-bit AD7578KN and ten-bit AD573JN are commercially available integrated forms of ADCs using the successive approximation techniques.

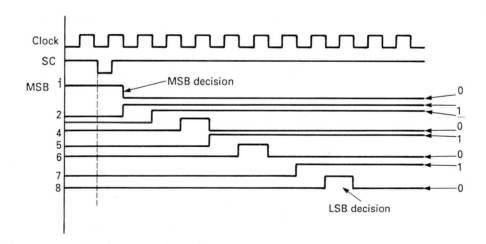

Figure 6.16

SINGLE- AND DUAL-SLOPE CONVERTERS

These kinds of converter use a ramp generator, integrator, comparator and binary counter. Examples are to be found in the digital voltmeter, where decoding circuits are added to drive a liquid crystal or LED display.

You will find an analysis of these systems in the section 'Methods of measurement' in *Electrical and Electronic Principles 3* (2nd ed), and we will not pursue them here.

1 The inverting input of an opamp is earthed and negative feedback is applied from output to the non-inverting input. What voltage would you measure at the non-inverting terminal?

2 What voltage would you find at the inverting input of an opamp with negative feedback if the non-inverting input was taken to −5 V?

3 Define the following terms: (a) binary-weighted network (b) bipolar DAC (c) quantization interval (d) follower converter (e) parallel converter.

4 A four-bit weighted-resistor DAC has an LSB resistor value of 100 kΩ. What are the values of the other resistors? How many output levels does this converter have?

5 For a DAC define the following: (a) quantization error (b) linearity error (c) offset error.

6 What is the major disadvantage of a weighted-resistor type of DAC? How is this overcome in the R-$2R$ ladder network?

7 Draw a diagram of a four-bit R-$2R$ DAC. In such a circuit, the next to MSB switch is closed and the other three switches are earthed. Draw an equivalent circuit for this condition.

7 Thyristors and Triacs

Aims: At the end of this unit section you should be able to:
Describe the physical structure and explain the operation of solid-state relays in the form of thyristors and triacs.
Determine the static characteristics of thyristors.
Describe the principles of phase control and integral cycle control systems.
Understand the use of thyristors in DC supplies.

The efficiency of any power amplifier depends upon the magnitude of that part of the energy available from the source which is wasted as heat within the amplifier. In class A operation, where the device (a bipolar or field-effect transistor) operates continuously on the linear part of the characteristic, the efficiency is low. An improvement follows by using class B operation, where the device is switched, being quiescent or off for roughly half the input cycle. Class C, in which the operation takes place over only one-third of the cycle, provides an even higher efficiency.

It appears, then, that to avoid power dissipation within a control system, the approach to study is the one where the device should be either fully conducting, that is switched hard on, or fully non-conducting, that is switched off. All intermediate conditions involving part conduction such as are found in class A systems must be avoided.

Look at the problem in the way illustrated in *Figure 7.1(a)*. Here a switch is connected in series with the energy source and the load. This switch constitutes the 'amplifier'. This may seem a ridiculous statement at the moment, but you should not think of this switch as being an ordinary mechanical toggle. We are controlling a heavy load by a *form* of switch which is itself controlled by a light-duty component. This is the whole principle of amplification. When the switch is open, no power is dissipated in the load and no power is dissipated in the switch since the current through it is zero. When the switch is closed, maximum power is dissipated in the load but again no power is dissipated in the switch since the voltage developed across it is zero. This assumes that we have an ideal switch of either zero

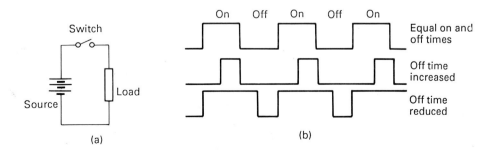

Figure 7.1

or infinite resistance; in real life this would not be so, but the value of the idea lies in the fact that power is wasted *only* during the periods when the switch is actually opening or closing. If this switching cycle is rapid enough, then the average power dissipated in the load will depend entirely upon the relative proportions (or the mark–space ratio) of the time for which the switch is open or closed; hence for an ideal switch, 100 per cent efficiency would be achieved.

Clearly, the flow of power to the load can be varied by altering the mark–space ratio or the *duty cycle* of the switch operation. In practice it is usual to vary the off time while keeping the sum of the successive on and off periods constant; this is illustrated in *Figure 7.1(b)*. The point to notice is that the pulse *repetition* frequency is unchanged. The choice of this frequency depends upon the application but, because it is only the average power which is being varied, the frequency has to be made high in relation to the most rapid rate at which the load power is required to vary. Looking at Figure 7.1 again, it should be clear that the fastest response to control signals can only, in the extreme, be half the repetition rate, because going from off to on would only correspond to half a cycle of control signal. Since the control signal is usually the AC mains supply, the repetition frequency is either 50 or 100 Hz depending on the circuit configuration.

SWITCHING DEVICES

We have already investigated diode and transistor switching circuits in *Electronics 2*. The diode is the most simple of all possible electronic switches; it is on whenever the applied voltage provides sufficient forward bias. The transistor also makes an unsophisticated form of switch; collector current flows whenever the base current is large enough to turn the switch on. Although it is widely used in digital switching systems where only a moderate level of power is controlled, the transistor (and the diode) has the disadvantage that there is an appreciable voltage drop across the device when it is switched on (about 0.2 V for the transistor, 0.6 V for the diode), and hence it is far from being an ideal switch in that there is high internal dissipation. Further, a high and continuous base current is required to keep the transistor switched on. Other devices have therefore been designed specifically to do the job of switched amplifiers in the sense that we are now using the term.

Example 1
A DC supply is switched so that the load current follows the waveform shown in *Figure 7.2*. What percentage of full power is applied to the load?

From the diagram the load current flows for 6 ms and is then off for 14 ms. The fraction of full power (which would be represented by a continually on waveform) is clearly 6/20 or 3/10. The percentage of full power delivered is therefore 30 per cent.

---

6 ms   14 ms

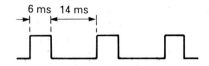

Figure 7.2

Figure 7.3

Example 2
The load current flowing from a half-wave rectifier is shown in *Figure 7.3*. What now is the percentage of full power delivered to the load?

We need caution here. We know that the 'average' value of this waveform is $1/\pi$ or 0.318, but this does not concern us in the present case. The *duty cycle* is clearly 1/2; hence the fraction of full power delivered to the load is 1/2 or 50 per cent. We can look at it this way: since the duty cycle is 1/2, the RMS value is $\sqrt{(1/2)}$ of the full sine wave, which is itself $\sqrt{(1/2)}$ of the peak value. Hence the RMS value is $\hat{I}/2$.

THE THYRISTOR

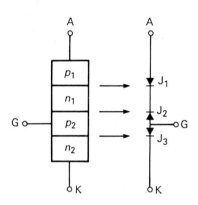

Figure 7.4

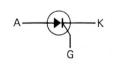

Figure 7.5

Suppose we have a four-layer semiconductor device made up of alternate layers of *n*- and *p*-type material, as shown in *Figure 7.4*. We might treat this arrangement as three *p-n* junctions in series, as indeed it is. By bringing out three terminations A,G and K, we get the device known as a *thyristor* or *silicon-controlled rectifier* (SCR). The terminals are known respectively as the anode, the gate and the cathode, and the circuit symbol for this device (which is a *p*-gate control) is shown in *Figure 7.5*.

Suppose that a voltage is applied between A and K, with A being negative. Then junction J_2 will be forward biased and junctions J_1 and J_3 will be reverse biased. Viewed overall, therefore, the three junctions behave as a reverse-biased assembly and no current can flow *unless* the applied voltage becomes great enough to exceed the avalanche breakdown points of junctions J_1 and J_3. The characteristic is therefore similar to that of an ordinary reverse-biased diode.

Suppose now that the applied voltage is reversed, with A being positive. This time junctions J_1 and J_3 are forward biased and J_2 is reverse biased. Again an overall view shows that the assembly is non-conducting but that only a *single* junction, J_2, is preventing the flow of forward current. A sufficient increase in the applied voltage could cause an avalanche breakdown in J_2, and we would then expect a large current to flow abruptly through the thyristor. When this happens there is a sharp drop in the voltage across it. The voltage–current characteristic for this on condition is quite unlike what we would expect from an ordinary forward-biased string of diodes. *Figure 7.6* shows a typical characteristic, and we will return to this shortly.

What actually happens can be best explained in a qualitative way by considering the layering of the junctions to be split up into a pair of interconnected transistors. This concept is illustrated in *Figure 7.7*. Assume now that the gate terminal is at zero volts or slightly negative. The *n-p-n* transistor of the pair (T_2) will be cut off and its collector current (neglecting a very small leakage current) will be zero. The collector current of T_2 provides base current for the *p-n-p* transistor T_1 (and vice

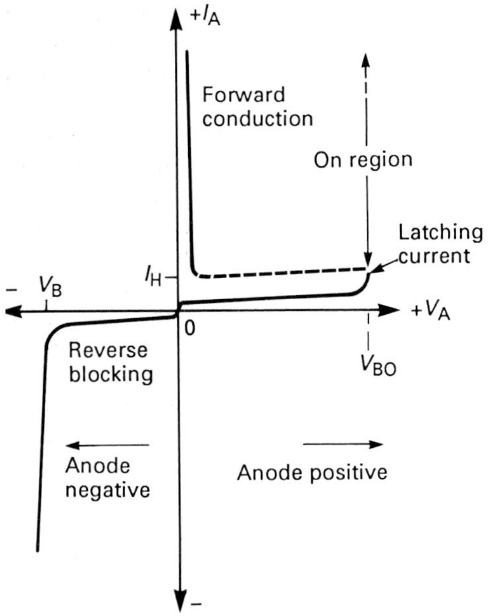

Figure 7.6

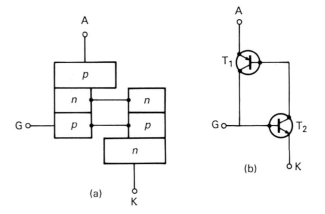

Figure 7.7

versa); hence if one transistor is turned off, the other must also be turned off. In this condition the thyristor as a whole is switched off or in a *blocking* state.

If now the gate terminal is made sufficiently positive, collector current will begin to flow in T_2. This provides a base current for T_1, which also begins to conduct. As the collector of T_1 now provides additional base current for T_2 a cumulative action is set up; in a very short time both transistors are driven into saturation. In this condition, the thyristor as a whole is switched on. Thus the gate input acts as a trigger which switches the device from off to on and does it very rapidly, typically in 1 μs.

Once the thyristor is switched on in this way, the gate loses its control over events; the gate input can be removed completely and the thyristor will remain switched on. The minimum forward current to maintain conduction *after* the removal of the gate current is called the *latching* current.

Generally, the collector current of T_1 is very much greater than the gate current; hence the gate input is ineffective in switching the thyristor off. The only method of doing this is to reduce the anode voltage so that the current falls below a critical level known as the *holding* level, I_H. This occurs when the anode voltage is reduced to zero or made negative. The characteristic of Figure 7.6 can now be explained. If the gate is maintained at zero or a small negative potential and the voltage on the anode is increased negatively, avalanche breakdown of the two outer junctions of the thyristor occurs at the point V_B. If the anode voltage is increased positively, the centre junction approaches avalanche breakdown, and the avalanche current will have the same polarity as a positive gate current. Hence, as breakdown nears, the thyristor 'snaps' itself on. The anode potential which will switch the device on, even though the gate current is zero, is known as the breakover voltage, V_{BO}.

Figure 7.8 shows a family of characteristic curves for a typical thyristor. The levels of the breakdown and breakover voltages depend upon the form the fabrication of the thyristor takes, but can range from 200 V to 2000 V. Current handling capacity can be up to as much as 1000 A.

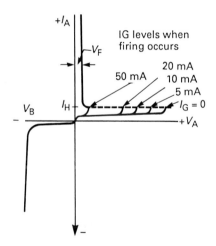

Figure 7.8

Each of the curves is plotted for a different value of gate current. As the gate current is increased, the anode voltage at which switching takes place is reduced, that is, the breakover level is reduced. The holding current I_H in this example is 50 mA, which is typical of small-power thyristors. For $I_G > 50$ mA, the thyristor is permanently on and behaves as an ordinary diode. The forward voltage drop V_F when the thyristor is on is usually very small, of the order of 1 V.

1 In the previous discussion, the base of the *n-p-n* transistor is used as the gate input and accepts a

positive-going gate signal. Is there any reason why the base of the *p-n-p* transistor should not be used as the gate input by accepting a negative-going gate signal?

We can now briefly look at thyristor action in a quantitative way by returning to Figure 7.7. Let the common-base current gains of T_1 and T_2 be respectively α_1 and α_2. Consider the current I crossing the collector junction of T_1. This current is made up of $\alpha_1 I$ (the hole current of T_1), $\alpha_2 I$ (the electron current of T_2), and the total leakage current I_{CBO}. Hence

$$I = \frac{I_{CBO}}{1 - (\alpha_1 + \alpha_2)}$$

From this result, if the *sum* of the two current gains is unity, the current will be infinite in theory, though limited in practice by the external circuit resistance. You should recall that the actual value of the α gain of a transistor depends upon the current flowing, being quite low at small values of emitter current but rising towards unity as the current increases. By proper design, it is possible to arrange for the sum of the αs at a very low forward bias to be just less than unity; the current will then be perhaps ten times I_{CBO}. This is still an extremely small current and the thyristor as a whole is an effective block to the passage of current.

If the forward bias is now increased, the breakover voltage is eventually reached and avalanche breakdown occurs in the centre junction, augmenting the gains α_1 and α_2. As $\alpha_1 + \alpha_2$ tends to unity, I becomes very large; in fact, the increase in current itself tends to increase $\alpha_1 + \alpha_2$ so that the avalanche multiplication need not be very great to sustain the condition that $\alpha_1 + \alpha_2 = 1$. The thyristor is now switched on. The locus of the characteristic between the breakover voltage V_{BO} and the holding current I_H (see Figure 7.6) is given by $\alpha_1 + \alpha_2 = 1$. By reducing the supply voltage, the current is reduced and a point is reached (I_H) where the current no longer sustains the α sum at unity; switch-off then occurs. The breakover voltage V_{BO} is reduced considerably on injecting current into the gate terminal, since the current flowing, and hence $\alpha_1 + \alpha_2$, is increased.

Values of α of 0.5 and less are, of course, much too low for ordinary discrete transistors, where 0.99 is the general norm.

Example 3
Suppose that the α dependence on the voltage and current conditions of the equivalent transistors built into a thyristor is given by

$$\alpha = 0.35 + \frac{V}{1000} + \frac{I}{50}$$

where I is in mA. Estimate the values of V_{BO} and I_H for this thyristor.

At the breakover point the voltage is high but the current is low. Therefore by ignoring the effect of current on the gain we have

$$0.5 = 0.35 + \frac{V_{BO}}{1000}$$

since the gains α_1 and α_2 for the two sections of the thyristor must each be 0.5. Hence $V_{BO} = 150$ V.

At the holding level the voltage will be low and most of the extra gain will be due to I_H. Hence

$$0.5 = 0.35 + \frac{I_H}{50}$$

from which $I_H = 7.5$ mA.

---

2 The figures used in Example 3 were quite hypothetical. Nevertheless, can you say which of the results for V_{BO} and I_H is likely to be the more accurate?

THE SWITCH AMPLIFIER

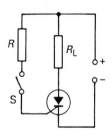

Figure 7.9

It is now possible to get an interpretation for the otherwise curious conception of a switch 'amplifier' which was introduced at the beginning of this chapter. The input to the 'amplifier' is the small gate current, a matter of perhaps 10 to 20 mA. The output is a very large current, possibly hundreds of amperes, flowing through some kind of load. *Figure 7.9* shows a very simple switch amplifier using a thyristor. When the switch S is open, the thyristor gate current is zero and the anode current is also zero. When the switch is momentarily closed, a gate current flows which is determined by resistor R, and the thyristor switches on. A large current then flows through the load R_L.

This, of course, is not a very practical circuit as there is no way of switching the thyristor off without removing the DC supply. We will return to this later on. The important point now is that only a momentary operation of a light current switch is enough to control a very large load current.

Example 4
Figure 7.10 shows a thyristor used to control a unidirectional load current. The input is the 240 V 50 Hz mains supply, and the gate is forward biased by a current which makes the breakover voltage V_{BO} to be 200 V. Sketch the output waveform of voltage across the load R_L, and discuss the limitations of this method of gate control.

On all negative half-cycles of the input waveform, the thyristor will be switched off and the output voltage will be zero. On the positive half-cycles, the thyristor will fire or switch on as soon as the anode voltage reaches 200 V or V_{BO}, and will remain switched on until the input half-cycle falls to zero. It will then remain off until the succeeding positive half-cycle reaches V_{BO}.

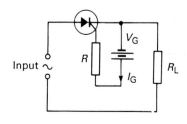

Figure 7.10

The output waveform will consequently be as illustrated in *Figure 7.11*. The angle represented by each full half-cycle is 180°. The thyristor therefore conducts over 144° or, as we say, the *conduction angle* θ is 144°. The angle of 36° is known as the *firing angle* φ.

This sort of circuit has the merit of simplicity. Its main disadvantage is that the thyristor can be fired only within the first 90° of the positive half-cycle, so we cannot obtain a conduction angle which is less than 90°. Notice the definitions of the terms 'firing angle' and 'conduction angle' from the diagram.

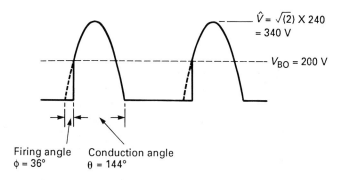

$\hat{V} = \sqrt{(2)} \times 240$
$= 340$ V

$V_{BO} = 200$ V

Firing angle
φ = 36°

Conduction angle
θ = 144°

Figure 7.11

FORMS OF THYRISTOR: THE TRIAC

If you attempted Problem 1 earlier on (and checked against the solution given) you will have realized that the *n*-gate-controlled thyristor, in which a negative gate signal is required for firing, is perfectly feasible. These forms of thyristor are sometimes known as anode-controlled devices, whereas the *p*-gate type are known as cathode-controlled devices. The symbol for an *n*-gate thyristor is shown in *Figure 7.12(a)*.

Thyristors are available in which the gate lead is brought out from both the *n*- and the *p*-type junction. This is known as the silicon-controlled switch (SCS) and its symbol is shown in *Figure 7.12(b)*.

The thyristors so far discussed are essentially half-wave devices since, if a sinusoidal input waveform is applied, they must reject the half-cycle which makes the anode negative, irrespective of whatever the gate signal is doing for the conducting half-cycle when the anode is positive. For full-wave applications, two thyristors are mounted in the same package

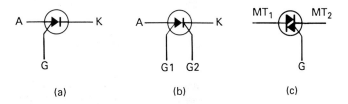

(a)

(b)

(c)

Figure 7.12

and called a *triac*. Here the anode of each device is connected internally to the cathode of the other. A single-gate electrode connected internally to both gate areas is provided. Triacs can be switched from an off state to an on state for either polarity of the applied anode voltage with positive or negative gate triggering. The symbol for a triac is given in *Figure 7.12(c)*. Because the triac will conduct in both directions, its main terminals are referred to as just that: MT_1 and MT_2. The triggering modes are always referred with respect to MT_1, and are designated as follows

1+	MT_2 positive, G positive
1−	MT_2 positive, G negative
111+	MT_2 negative, G positive
111−	MT_2 negative, G negative

APPLICATIONS

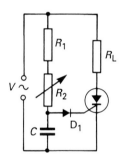

Figure 7.13

The thyristor in its various forms finds the widest applications in power control of AC systems, since the input voltage continually reverses in polarity, so allowing the thyristor to turn off. We have already discussed a very elementary form of control in connection with Figure 7.10, and a rather more advanced form is shown in *Figure 7.13*. Here the input is AC and the gate signal derived from this controls the fraction of each forward half-cycle, that is, the phase angle over which conduction takes place. The load R_L is usually a lamp or small electric motor; the circuit thus operates as a dimmer system or a motor speed control. Control of the gate is managed by the *RC* circuit $(R_1 + R_2)C$.

Assuming that the gate current is negligible, the current flowing through capacitor *C* will be

$$I_C = \frac{V}{\surd[(R_1 + R_2)^2 + X_C^2]}$$

The voltage across *C* is

$$V_C = I_C X_C = \frac{V}{\surd[(R_1 + R_2)^2 + X_C^2]} X_C$$

$$= \frac{V}{\surd\{[(R_1 + R_2)/X_C]^2 + 1\}}$$

Now the magnitude of V_C relative to V depends upon whether $[(R_1 + R_2)/X_C]^2$ is smaller or greater than unity. If it is very small relative to unity, V_C will be roughly equal to V; on the other hand, if it is very large compared with unity, V_C will be approximately equal to

$$\frac{V}{\surd[(R_1 + R_2)/X_C]^2} = \frac{V}{R_1 + R_2} X_C$$

This value of V_C will lag V by approximately 90° and will have a magnitude much smaller than V. Hence, for a given value of X_C, both the phase angle and the magnitude of the gate voltage ($= V_C$) can be varied by adjustment of the resistance. Only one resistor need be varied; the other acts as protection against excessive gate dissipation.

> 3 What purpose do you think diode D_1 serves in the circuit of Figure 7.13? Would the circuit work without it?
> 4 Assume the capacitor C to be removed from the circuit. Why can the phase angle at which the thyristor fires extend only between 0 and 90°?

We can now examine how the magnitude and phase angle of the gate voltage can control the time within the positive half-cycle of input when the thyristor fires.

First of all, suppose that the resistance chain $R_1 + R_2$ is very small, so that $[(R_1 + R2)/X_C]^2$ is also small. Then the gate voltage will have practically the same magnitude and phase as the AC input. As a result, the gate voltage exceeds the firing potential almost at the start of the positive half-cycle and the thyristor conducts throughout the half-cycle, simulating the effect of an ordinary diode. When the input passes through zero into the negative half-cycle, the thyristor switches off and the gate regains control. The cycle repeats when the anode again goes positive. *Figure 7.14(a)* shows the input waveform together with the gate potential ($=V_C$) and the load current.

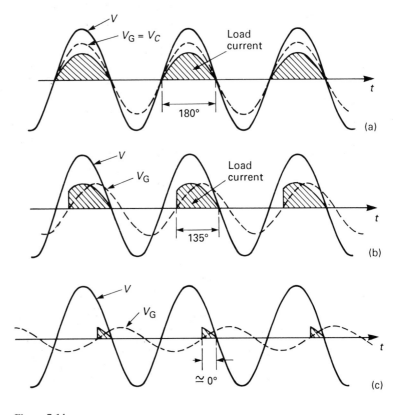

Figure 7.14

Suppose now that R_2 is increased in value sufficiently to make $[(R_1 + R_2)/X_C]^2 = 1$. Then $V_C = V/\sqrt{2}$ and the gate voltage will lag V by 45°. The thyristor will now fire only when the anode voltage has reached an angle of 45° into the positive half-cycle. This situation is illustrated in *Figure 7.14(b)*. Current now flows through the load for only three-quarters of the half-cycle; hence the power supplied to the load is reduced. Notice also that the gate potential amplitude is reduced.

Now suppose R_2 to be further increased until $[(R_1 + R_2)/X_C]^2$ is very much greater than 1. The gate potential now lags the input voltage by an angle approaching 90°. Further, the gate potential amplitude is now very small. The thyristor will consequently not fire until the gate voltage is almost at its maximum value. The conduction angle is therefore very small and the thyristor switches off almost immediately after switching on. *Figure 7.14(c)* illustrates this situation.

Thus phase control (as it is called) is the process of rapid switching which connects a supply to a load for a controlled fraction of each half-cycle (or cycle), the fraction being determined by phase. It has the advantage that virtually no power is dissipated as heat, because we are not *regulating* current as we do with a rheostat, but switching it. A thyristor is always fully on or fully off, and in either case negligible power is dissipated within it. Its disadvantage is that the maximum output is limited to a half-wave, but this can be overcome by the use of the triac.

We have already noted that the triac is a bidirectional thyristor which can be considered as a pair of inverse-paralleled thyristors under the control of a single gate terminal. The device can be switched on by either a positive or a negative gate pulse, irrespective of the polarity of the AC supply at that time.

If the triac is applied to the control of a lamp or motor load as covered previously in Figure 7.13, the circuit can be modified to that shown in *Figure 7.15*. Here the load current can be controlled from zero to full conduction by adjustment of R_2. The voltage and current waveforms for this system, assuming a conducting angle of 90°, are shown in *Figure 7.16*. This should be compared with the waveforms for the half-wave thyristor illustrated earlier.

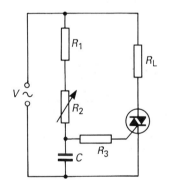

Figure 7.15

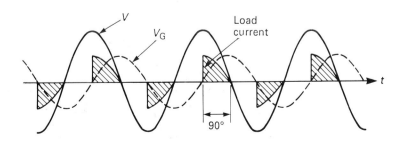

Figure 7.16

AN EXPERIMENTAL CIRCUIT

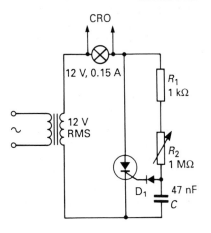

Figure 7.17

A simple experiment to show the working of a thyristor as a lamp dimmer can be made using the circuit of *Figure 7.17*. A suitable device is the IN5062 which has a forward current capability of 0.8 A and a reverse voltage rating of 100 V, although any low-power thyristor having similar parameters may be employed. It need not be fitted to a heat sink, as a low AC supply is derived from a transformer and the load is a 12 V 0.15 A bulb. Two 6 V bulbs may be used in series if desired.

Set the potentiometer R_2 (1 MΩ) to its minimum value and connect a CRO across the bulb terminals. A slow timebase should be selected as the output will be at 50 Hz frequency. Switch on and progressively increase the resistance of R_2, noting the effect on the brightness of the bulb and the displayed waveform. It should be possible with the circuit values given to just reduce the conduction angle to zero and hence extinguish the bulb. Why can the phase angle at which the thyristor is switched on only be varied between 0 and 180°?

You may find you need to experiment with the value of capacitor C if you use an alternative thyristor.

Now get hold of a triac and repeat this experiment using the circuit of Figure 7.15 as your guide, replacing R_L with the bulb and using a 12 V RMS supply.

INDUCTIVE LOADS

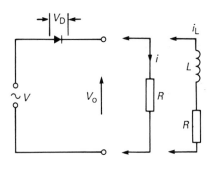

Figure 7.18

We have so far assumed that the loads connected to the thyristor-controlled circuits have all been purely resistive. This is a very unlikely circumstance in practice, particularly where the speed control of motors is concerned. It is informative first to look at the way an ordinary diode rectifier behaves when it feeds into an inductive load.

Figure 7.18 shows a simple half-wave rectifier feeding either a load which is purely resistive, or a load having some inductive reactance. The waveforms developed across the load and the load current are familiar to us. For the resistive load, these waveforms are as illustrated in *Figure 7.19(a)*. The turn-on and turn-off times of the diode are so small relative to the period of a 50 Hz input wave (20 ms) that their effect can be ignored. When the diode is switched off, the whole of the negative half-cycle is developed across it.

Now suppose the resistive load is replaced by a resistive-inductive combination. As for the previous case, current flow will begin as soon as the supply voltage goes positive, but the presence of inductance will delay the *change* in current. Hence at the end of the positive half-cycle the current will continue to flow for a time, the diode will remain switched on, and the load will look into the negative supply voltage until the current falls to zero. This is illustrated in *Figure 7.19(b)*. The average load voltage is now less than it was in the case of the resistive load.

It is clear, then, that when a thyristor is used as a half-wave controller, the waveforms we sketched for a resistive load (Figure 7.14) will no longer apply to a part-inductive load. The thyristor will only conduct, as we know, when its anode voltage is positive and its gate has received a firing signal. During that period when the thyristor is now switched on, the load current is dictated by the relative values of R and L in the load (and the rate of change of the current), but once the load voltage reverses then the load current begins an exponential decay. If this decay carries the current level below I_H for that particular thyristor, then the load current becomes discontinuous. This will tend to happen more as the firing angle becomes greater, since the mean load voltage is then lower. The waveforms now are shown in *Figure 7.19(c)*.

The problem can be overcome by connecting a diode in parallel with the load. This diode application goes under a variety of names, but it is best described as a *commuting* diode as its purpose is to transfer or 'commute' load current away from the thyristor whenever the load voltage goes into reverse. It serves two functions: it prevents a reversal of load voltage, and it allows the thyristor to reset to its blocking state at the point of zero voltage by transferring the total current away from the thyristor.

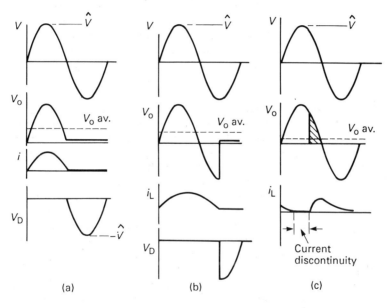

Figure 7.19

Example 5
A thyristor has a latching current of 40 mA and is fired by a pulse of 100 μs duration. It feeds an inductive load made up of a coil of 1 H having a resistance of 40 Ω from a DC source of 120 V. It is found that the thyristor does not remain on when the firing pulse ends. Explain why this is so, and suggest a method of overcoming this problem.

We can solve things mathematically here. As soon as the voltage is applied to the load, the current will rise exponentially towards its maximum possible value of V/R or $120/40 = 30$ A (*see Figure 7.20*). Since the time constant of L/R is long compared with the pulse length (0.025 s versus 0.0001 s) we can assume that the rise in current is

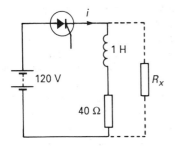

Figure 7.20

linear over the 100 μs period. Hence the rate of rise of current is V/L or 120/1 = 120 A/s. Therefore in 100 μs the current will be 12 mA. But this level is below the latch level of 40 mA; hence the thyristor will not remain switched on when the pulse ends.

By connecting a resistor in parallel with the load, a current of 40 mA can be made to flow immediately at switch-on. Let this resistance be R_x. Then

$$R_x = \frac{120}{(40 - 12) \times 10^{-3}} = 4.28 \text{ k}\Omega$$

A preferred value of 3.9 kΩ could be used here.

INTEGRAL CYCLE CONTROL

This is an alternative method of thyristor control to that of phase control. It has certain advantages for particular applications. It also avoids the generation of radio-frequency interference, which turns up in thyristor and triac control systems because of the rapidity of switching and the heavy currents involved. When the load currents are very heavy, such as in furnace control, the usual interference suppression methods are neither very effective nor economical. The intregral control method (or *burst firing* as it is often called) overcomes the problem by having the thyristor switched on for an integral number of half-cycles, then turned off for an approximately similar period. Both on and off switching actions take place at the moment when the input is passing through the zero voltage point. *Figure 7.21* shows the principle of the method. The gate firing pulses are electronically generated and controlled so that the zero voltage crossings on the input wave coincide with the switching points.

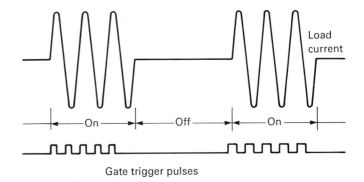

Figure 7.21

The method is used where heating loads are concerned. Switching every cycle is unnecessary because of the (usually) long thermal time constants of such loads. Little variation of the heater temperature will occur if the input to the heating element consists of a number of cycles on and a number of cycles off, as opposed to switching every cycle.

DC OPERATION   For control of DC supplies, the ordinary transistor switch, in
which the on and off state is under the control of a relatively
small base current, has an advantage over the thyristor in that
the latter has to have its anode voltage reduced to zero in order
to be switched off. In AC supply systems, the thyristor turns off
when the current naturally reaches its zero level in the AC
cycle. For this reason, ordinary transistors tend to be used for
low-power applications in preference to the thyristor. However,
such systems as battery-operated flashers, automatic car-parking
lights, burglar alarms and miniature drill speed controls
all involve DC operation.

The simplest way of turning a thyristor off is to incorporate a
push-button switch in series with the anode feed. The switch is
normally closed, and is opened whenever the thyristor needs to
be reset. This method, of course, takes the shine off the idea of
thyristor control because we are using a switch to interrupt the
whole of the load current (which flows via the thyristor), and we
might just as well retain the switch to start the load current up
as well.

A simple method of switching the device off is to connect a
capacitor in parallel with it; this capacitor carries a charge with
the polarity shown in *Figure 7.22*. When the thyristor is on, the
current flowing will be V/R_L. If now the capacitor is brought
into circuit by the operation of switch S, a reverse voltage will
appear across the thyristor and switch-off will occur. From the
waveform of the thyristor voltage V_T, the capacitor will recharge
through the load R_L as time elapses and will become oppositely
charged to the level of the applied voltage V. It is necessary for
proper operation that the capacitance is large enough to
maintain a reverse voltage across the thyristor for the required
turn-off time.

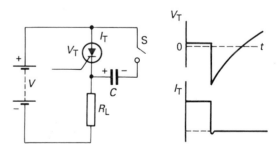

Figure 7.22

The switch S must, of course, be electronically controlled in
some way, and the capacitor C must be recharged to its original
polarity in time for the next switch-off sequence. One way of
doing this is shown in the circuit of *Figure 7.23*. Here a second
thyristor is used to control the first. The turn-on of the load
current is effected by a momentary pulse applied to the gate of
T_1. T_1 then fires and capacitor C starts to charge by way of R_1
and the short-circuit presented by T_1. Point X is hence at zero
potential, and point Y approaches the supply level V. If now a
momentary pulse is applied to the gate of T_2, T_2 will turn on

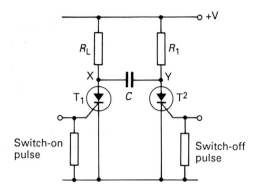

Figure 7.23

and the potential at Y will fall to zero. Since the charge on the capacitor cannot change instantaneously, the voltage cannot change either; hence point X must fall to a potential equal to $-V$. This will have the effect of turning T_1 off, after the manner described above. The cycle will repeat on receipt of the following control pulses.

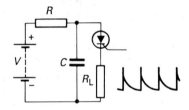

Figure 7.24

Example 6
A thyristor circuit shown in *Figure 7.24* has a fixed gate potential such that the device fires when the voltage across C has reached 100 V, and switches off when the voltage has fallen to 50 V. What sort of waveform appears across the thyristor load R_L? If this load value is 120 Ω and C is 10 μF, what power is dissipated in R_L for each cycle of operation?

This circuit produces a sawtooth waveform across the load represented by the discharge curve of C as its voltage falls from 100 V to 50 V. The energy stored in the capacitor during the charge (via resistor R) is dissipated in R_L during the discharge cycle.
   Ignoring losses in the thyristor, the energy released by C in discharging from 100 V to 50 V is $(100^2 - 50^2)C/2$ J (since energy stored is $CV^2/2$). So

$$P = \frac{100^2 - 50^2}{2t} C \ \text{W}$$

where t is the time of discharge. This can be found from the exponential equation for the discharge of a capacitor:

$$V_C = V\exp(-t/R_L C)$$
$$50 = 100\exp(-t/120 \times 10^{-5})$$

from which we get

$$833t = \log_e 2$$
$$t = 0.83 \ \text{ms}$$

Substituting values into the power equation, we have

$$P = \frac{100^2 - 50^2}{2 \times 0.83 \times 10^{-3}} \times 10^{-5}$$

$$= 45.2 \text{ W}$$

We shall be examining triacs and other forms of solid-state switching devices in the following section in more detail. You should try the following problems before going any further.

5 List the advantages (and disadvantages) of thyristors over (a) *p-n* diodes (b) transistors.

6 Explain the behaviour of a thyristor as breakover is approached and reached. Define the terms (a) reverse breakdown (b) breakover voltage (c) holding current (d) latching current (e) firing angle (f) conduction angle.

7 Why doesn't the removal of gate current turn off a thyristor? Under what conditions could this in fact happen?

8 What is the basic difference between the function of a thyristor and that of a triac?

9 Describe the principles of (a) phase control (b) integral cycle control as applied to thyristor circuits. Why wouldn't integral cycle control be suitable as a lamp dimmer system?

10 Looking back at Figure 7.14(b), suppose the peak voltage to be 200 V and the conduction angle 120°. If the load resistance is 10 Ω, find (a) the average load current (b) the power dissipated.

8 Triggering devices

Aims: At the end of this unit section you should be able to:
Understand the construction and operation of a unijunction transistor.
Use the unijunction as a relaxation oscillator and as a triggering device for thyristors.
Explain the operation of a diac.
Understand the operation of a gate turn-off thyristor.
Experiment with diacs and thyristors in practical circuits.

We have seen in the previous section how the silicon-controlled rectifier or thyristor is used in basic switching and control circuit systems. There are a number of other semiconductor devices allied to the thyristor; these are the unijunction transistor, the diac and the gate turn-off thyristor. The first two of these are essentially triggering devices. We will briefly examine the construction, the characteristics and the applications of these devices in this chapter.

THE UNIJUNCTION TRANSISTOR

The so-called unijunction transistor (UJT) is a versatile device which not only is suitable for firing thyristor and other control circuits but also has applications to timing circuits and clock pulse generation. Like an ordinary bipolar transistor it has three terminals, but unlike the transistor it has, as its name implies, only one internal semiconductor junction. Its physical construction is illustrated in *Figure 8.1*, together with its theoretical circuit symbol. A narrow bar of lightly doped *n*-type semiconductor has an off-centre single *p-n* junction. The connection taken from the *p*-region of this junction is known as the emitter. Connections are also taken from each end of the bar and known respectively as base 1 and base 2. These contracts are *ohmic* contacts – that is, they are non-rectifying.

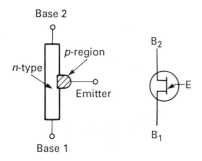

Figure 8.1

The way the UJT operates can be followed with the assistance of an equivalent circuit, and this is given in *Figure 8.2*. The silicon bar acts as a resistor (as it does in the FET) which is effectively tapped into by a diode where the junction occurs. Suppose we bias the various terminals of the unijunction in the manner shown on the right of the diagram, including a load resistor in the lead to B_2. Then a current will flow through the bar resistances R_{B1} and R_{B2}, and the potential at the cathode of the diode will be

$$V_J = \frac{R_{B1}}{R_{B1} + R_{B2}} V_{BB}$$

which will be some fraction η of the voltage on B_2 with respect to B_1. Hence

$$V_J = \eta V_{BB}$$

where η is known as the *intrinsic stand-off ratio*.

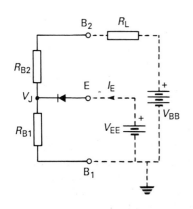

Figure 8.2

If the emitter bias V_{EE} is less than V_J, the junction will be reverse biased and only the leakage current will flow in the emitter circuit. However, if V_{EE} is increased to $V_J + 0.6$ V, the junction will become forward biased and a current will flow from the emitter to B_1. This happens because the injected carriers are holes; these will be repelled by the positive B_2 end of the bar, and attracted to the negative B_1 end. This flow through the equivalent resistor R_{B1} results in a reduction in its resistance, which in turn reduces the emitter voltage. Thus we have a condition where the emitter current is increasing but the emitter voltage is decreasing; this is a condition of negative resistance. Pure negative resistance does not exist, of course; you cannot go into a shop and buy one off the shelf. But there are a number of devices, like the unijunction, where a graph of current against voltage has a negative gradient. A typical current–voltage characteristic for a unijunction transistor is shown in *Figure 8.3*. The peak point on the curve, V_P, is that voltage at which the unijunction switches on, and here $V_P \simeq \eta V_{BB} + 0.6$ V. As more holes are injected, saturation is eventually reached and the emitter voltage begins to increase again with increasing emitter current.

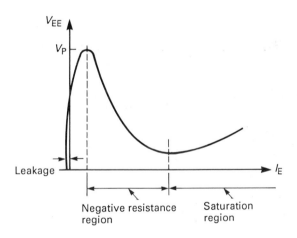

Figure 8.3

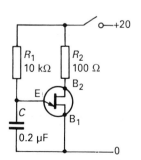

Figure 8.4

Example 1
The 2N2646 unijunction transistor has $R_{B1} + R_{B2} = 7$ kΩ and $\eta = 0.65$ as typical values. It is used in the circuit shown in *Figure 8.4*. Show that the output across capacitor C will be a sawtooth wave, and estimate the period of this wave.

For the first part, suppose that at time $t = 0$ the voltage across C is zero and the emitter junction is consequently reverse biased. The unijunction is therefore off. Capacitor C will charge up through resistor R_1 until at some time t_1 later the voltage across it will equal the peak voltage V_p. At this point the transistor switches on and the emitter-to-B_1 voltage falls, permitting C to discharge. This turns the transistor off again at time t_2 and the charging cycle begins afresh. The voltage waveform developed across C is therefore a sawtooth, as shown in *Figure 8.5*.

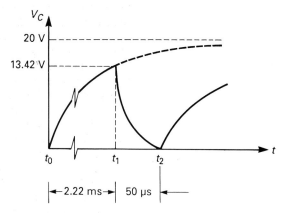

Figure 8.5

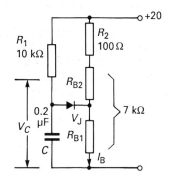

Figure 8.6

To establish the period of this waveform we use *Figure 8.6*, beginning at time t_0 when C is uncharged and the emitter potential is 0 V. As soon as the circuit is switched on, current flows through R_2, the unijunction transistor B_2 to B_1, and back to the supply. This current will be

$$I_B = \frac{V_{BB}}{R_2 + R_{B1} + R_{B2}} = \frac{20}{7100} \text{ A} = 2.82 \text{ mA}$$

The voltage on B_2 is

$$V_{BB} - I_B R_2 = 20 - (2.82 \times 10^{-3})100 = 19.72 \text{ V}$$

Call this V_{B2}. Then the voltage at the junction point V_J will be

$$\eta V_{B2} = 0.65 \times 19.72 = 12.82 \text{ V}$$

The capacitor will charge, therefore, from 0 V towards 20 V with a time constant of
$CR = 0.2 \times 10^{-6} \times 10^4 = 2 \times 10^{-3}$ s or 2 ms. The equation of the charge is therefore

$$V_C = 20(1 - \exp -t/2 \times 10^{-3})$$

The charge will continue until the voltage has risen to 12.82 + 0.6 V when the transistor will switch on. From Figure 8.5, therefore,

$$13.42 = 20(1 - \exp -t/2 \times 10^{-3})$$

from which we find $t = 2.22$ ms.
    This is the period of the wave if we ignore the time of discharge. We will look at this question in what follows.

It is of interest to see what happens at B_2 during the charge and discharge cycle; what sort of output do we get there? We can answer this by analysing the discharge condition in rather more detail than we did a short time ago. When the capacitor voltage reaches 13.42 V (using Example 1), the emitter junction becomes forward biased and hole carriers are injected into the base region; these move towards the negative (earth) end of the bar and reduce the resistance of section R_{B1}. As we have already indicated, this reduces the effective resistance of R_{B1} with a consequent drop in the voltage at the junction point. This action causes the junction diode to become even more heavily forward biased (think about it), with the result that more hole carriers pass across into the base region, reducing the resistance of R_{B1} even further. This action is self-supportive and the ohmic value of R_{B1} falls very rapidly to its minimum, typically 20–50 ohms. Hence a very low resistance appears across the capacitor terminals immediately following 2.22 ms of the charging cycle. The time constant of the discharge curve, taking the minimum value of R_{B1} to be 50 Ω, will consequently be $0.2 \times 10^{-6} \times 50$ s = 10 μs, and the discharge will be completed (in a practical sense) in about 50 μs. Figure 8.5 illustrates this, though the scale of the discharge curve is greatly exaggerated for clarity.

Since the emitter junction can be looked on as a short-circuit during the discharge cycle, the bottom end of R_{B2} will be practically at zero volts during this period. So the bottom end of R_{B2} will drop from 12.82 V (while C is charging) to, say, 0.1 V while C is discharging. The top end of R_{B2} (terminal B_2) will similarly drop from 19.72 V to a much lower figure. (See if you can estimate it.) The output at terminal B_2 is consequently rather as shown in *Figure 8.7*, where the sawtooth across C is also drawn for comparison. By placing a resistor in the B_1 lead to earth, a trigger pulse can be obtained from this point also.

Such pulses as these can be used as triggers to operate thyristors or triacs in power control systems.

V_C

V_{B2}

Figure 8.7

Example 2
In the sawtooth oscillator circuit of Figure 8.4 it is found that if the value of R_1 is too large or too small, the circuit fails to operate. Explain why this happens.

Suppose R_1 is very large, of the order of a megohm or so. Then the current which would go to charge the capacitor may be comparable only with the normal leakage current of the diode junction. The charge on C would therefore reach only to some low level and remain there, unable to initiate the switch-on cycle. If C was an electrolytic having its own small leakage, the problem would be aggravated.

If R_1 is very small, of the order of a few hundred ohms, the current flowing through it and the diode junction after the circuit has fired once might be sufficient to keep the diode in the saturated region of its characteristic. The circuit would then quite clearly fail to oscillate.

Figure 8.8 shows a simple UJT oscillator used to trigger a thyristor in a lamp dimmer system. Although it is perfectly possible to trigger the thyristor directly from the B_1 terminal by

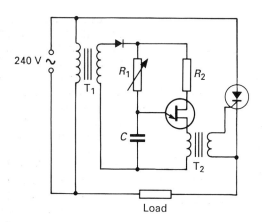

Figure 8.8

including a resistor in this lead, the use of an isolating pulse transformer such as T_2 separates the high-voltage mains supply from the trigger circuitry which can be operated from a low voltage supplied by transformer T_1. This is a quite common technique; often opto-isolators are used instead of pulse transformers.

THE DIAC A diac (or diode alternating current device) is a silicon bidirectional device suitable for firing thyristors. It is essentially a gateless triac; in fact its circuit symbol, shown in *Figure 8.9*, is seen to be equivalent to a triac without the gate terminal. It has two terminals, anode A_1 and anode A_2, and acts as two diodes in parallel, turning on only when the breakover voltage (strictly zener breakdown here) in *either* direction is exceeded.

The structure of a diac is shown in *Figure 8.10(a)* and, in spite of what has been said above, it is not simply two paralleled diodes. Five layers of semiconductor material are involved. However, if we imagine the structure split down the middle as depicted in *Figure 8.10*(b), we may treat it as a pair of electrically separate but physically connected four-layer devices in parallel, each reversed with respect to the other. This explains the form of the circuit symbol.

Figure 8.9

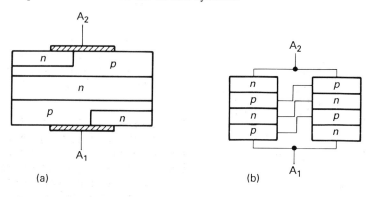

(a)   (b)

Figure 8.10

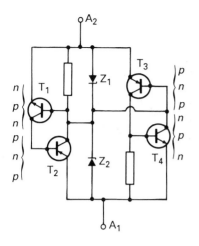

Figure 8.11

Suppose anode A_2 is made positive with respect to anode A_1, and the voltage is increased. It will help to understand the operation of the diac if we visualize Figure 8.10(b) as a discrete transistor circuit, as shown in *Figure 8.11*. What we have in essence is a four-layer arrangement making up two transistors on each side (which is essentially the assembly of a thyristor), with two zener diodes in the gating circuits. These zeners are themselves part of the system, so you must not look for them as discrete entities in the actual diac construction. When A_2 goes positive with respect to A_1 the current will initially be very small, but as the voltage is increased the breakover point of zener Z_2 will be reached. When this occurs, base current will flow from T_3 through Z_2. T_3 then begins to turn on, so that its collector current increases and supplies drive to the base of T_4. T_4 in turn now switches on and applies more base current to T_3. This effect is regenerative and, in a very short time, T_3 and T_4 are switched hard on. The resistance between A_2 and A_1 is then very small. This state of conduction cannot be removed until the applied voltage falls below the holding level. When the polarity between A_2 and A_1 is reversed, T_1 and T_2 form the conducting path in place of T_3 and T_4.

So we have a device which, looked at from either direction, remains off until the applied voltage reaches a precise level, when it then switches hard on. These switching characteristics are shown in *Figure 8.12*.

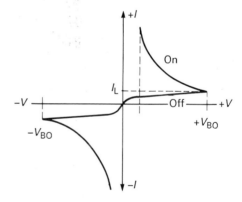

Figure 8.12

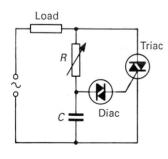

Figure 8.13

1 In an examination a student wrote: 'A diac is a triac with two added zeners.' Do you think this is a reasonable description of a diac? Give your reasons.

One common use for a diac is shown in *Figure 8.13*, where it controls the input to the gate of a triac. As there is a spread in the gate turn-on currents in thyristors, the diac is introduced into the gate feed path to ensure that the triac conducts at the same value of supply voltage each time. The diac, being bidirectional, detects both polarities and provides a voltage threshold that must be overcome before gate current can flow.

The diac fires at a precisely determined phase, controlled, as we have already seen, by the *CR* phase-shifting combination. On firing, the voltage built up on the capacitor is passed by the diac and the triac is switched.

THE GATE TURN-OFF THYRISTOR

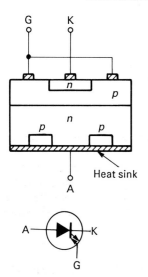

Figure 8.14

The gate turn-off thyristor (GTO) combines the high current and high blocking voltage of a conventional thyristor with the speed and ease of drive of an ordinary transistor. Like the thyristors we have already met the GTO is a four-layer device, and its general structure and circuit symbol are shown in *Figure 8.14*.

The GTO is turned on in the usual way by the application of a positive gate pulse but, unlike the conventional thyristor, it is turned off when a negative pulse is applied. Typically, a pulse length of only a few microseconds and of as little as -5 V amplitude is sufficient to effect turn-off.

The forward characteristics of a GTO are interesting. A study of a typical family of characteristics shows that there is a change in the form of the curves on either side of the current value which represents the latching level I_L. When the anode current is less than I_L, the device behaves as a high-voltage transistor; anode current rises with increasing anode voltage and then remains substantially constant, its actual level depending upon the gate current. Above the level of I_L, the characteristic moves into the thyristor form. If the gate current is less than that required for triggering, I_{GT}, the GTO is off and only a very small leakage current flows between anode and cathode. If the gate current is equal to or greater than I_{GT}, the GTO switches on, a large anode current flows and the voltage drop between anode and cathode is small. As long as the anode current remains below the latching level, however, the GTO will return to the off state if the gate current drops below I_{GT}.

Drive Systems

As the GTO operates with very short switching times and requires a polarity change at the gate for switch-on and switch-off respectively, different drive circuits to those used with ordinary thyristors are necessary.

To turn the GTO on requires a positive current to be injected into the gate for the necessary turn-on time. To turn the GTO off requires a negative voltage of between -5 and -10 V (typically) applied directly between gate and cathode for several microseconds, thus removing current from the gate. The actual voltage must be less than the gate–cathode reverse breakdown level, but high enough to remove the charge necessary to effect the turn-off.

Drive circuits usually operate to provide the turn-off pulse from a capacitor which has been allowed to charge up during the time that the GTO has been conducting; the actual amplitude of the pulse is determined by a voltage regulator (zener) diode. *Figure 8.15* shows a basic arrangment. T_1 can be switched on and off alternately by the incoming rectangular waveform. When T_1 is off, a positive drive current flows into the GTO gate by way of R_2 and C_1, with the initial value being set by R_2. The zener diode conducts as soon as its breakdown voltage is reached, thus holding the charge on C_1 to that level.

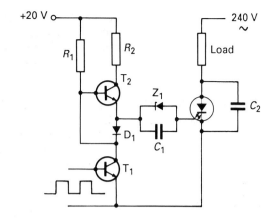

Figure 8.15

A small and necessary current then flows into the gate during the conduction period.

When T_1 switches on, T_2 switches off. Capacitor C_1 now discharges via T_1 and D_1, removing the gate current and turning the GTO off. The capacitor C_2 across the GTO is known as a snubber, and limits the rate at which the anode–cathode voltage rises at turn-off.

ZERO-VOLTAGE DETECTOR

Mechanical relays have always given problems after a period of use. Their contacts arc in spite of various methods of reducing this form of deterioration, and eventually contact becomes erratic or fails altogether. Solid-state relays of the kinds we have been discussing overcome this problem since there is no contact wear inside a semiconductor. Many solid-state relays have an internal *zero-voltage switch* (or detector) for firing the device in burst firing or proportional control applications. Zero-voltage switches are also available for the control of an external thyristor or triac.

In ordinary phase control, wh'ch we have discussed earlier, the power in the load is controlled by delaying the turn-on of current. However, this results in the production of radio-frequency interference which cannot always be fully suppressed by *LC* filter systems in the supply leads. With zero-point firing, the current is always switched on at the zero-crossing points of the AC input cycle, thus minimizing the generation of interference. High immunity against spurious triac firing under noisy mains conditions is also provided.

A block diagram of a typical zero-voltage switch is shown in *Figure 8.16*. The AC input, which is current limited, is rectified (and externally smoothed) and then fed to a regulator which provides a stable voltage for a control potentiometer *R*. The input also feeds to a period pulse generator and the zero-crossing detector. The output from the period pulse generator is a short-duration pulse for each completed mains input cycle. These pulses are used to act as logical clock signals so that, in burst firing control, only *complete* mains cycles are applied to the load. They are also used to control a ramp

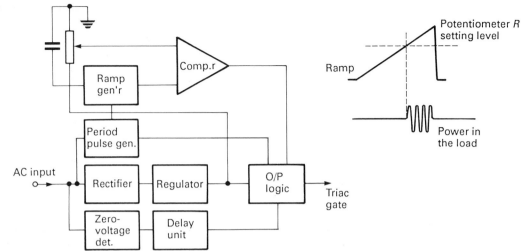

Figure 8.16

generator output whose duration is not dependent upon a large value for capacitor C.

A comparator takes inputs from the control potentiometer and from the ramp generator. Its output controls the logic circuitry, and the setting of the potentiometer defines the fraction of the ramp period for which the triac is in conduction, so controlling the power in the load. *Figure 8.17* shows how adjustment of the control potentiometer affects the duration of the load current. If the potentiometer is linear, the percentage power delivered to the load is also linear with rotation.

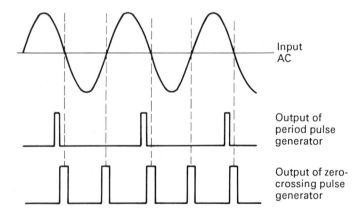

Figure 8.17

Zero-voltage switches are at their best when used as proportional controllers for temperature control in conjunction with a sensor such as a thermistor. The ramp period in such applications should be much shorter than the thermal time constant of the system but long enough to contain a large number of mains cycles.

A circuit using the SL443A zero-voltage switch is shown in *Figure 8.18* with suitable external components. The control potentiometer is R_2, while R_1 provides a reference at the input to the comparator.

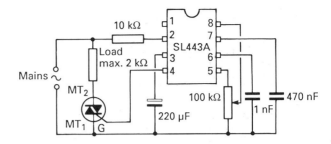

Figure 8.18

SUMMARY We have now worked our way through a number of switching devices. In practice, the actual device used will depend almost solely upon the application. All the devices mentioned have relative merits which make them more suitable for one particular application than another.

Although we have not included it in this survey, the conventional transistor, particularly the power MOSFET, is superior as a switch in terms of switching speed, with the bipolar power transistor a close second. Below switching speed requirements of some 20 kHz, and particularly at mains frequency, the thyristor family take over in terms of robustness, efficiency and overload capabilities. Transistors have the advantage in that they can operate at temperatures up to some 200°C, whereas thyristors are limited to about 125°C. It is easier, however, to provide protection against faults for the thyristor family than for transistors.

AN EXPERIMENT This experiment enables the intrinsic stand-off ratio η of a unijunction transistor to be determined, and its behaviour as a sawtooth generator to be looked at.

Using the 2N2646 UJT, make up the circuit of *Figure 8.19(a)*.

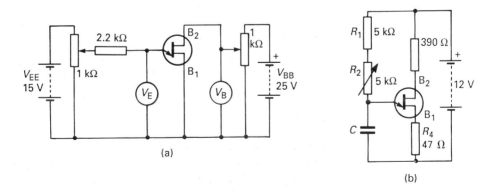

Figure 8.19

Power supply units may be used in place of the battery supplies indicated and, if these are adjustable, the potentiometers shown may be omitted. The voltmeters V_E and V_B measure respectively the emitter potential and the base 2 potential, and both meters should be high-resistance types. For a series of values of V_{BB}, the voltage at the emitter is slowly increased from zero until there is an abrupt drop in V_E, indicating that at this point $V_E = V_P$, the peak voltage. Select about five values of V_{BB}, such as 5, 10, 12, 15 and 20 V, and record the value obtained for V_P for each of them. Since

$$V_P = \eta V_{BB} + V_D$$

a graph can be plotted of V_P (vertically) against V_{BB} (horizontally). Values for η and V_D (the diode voltage drop, assumed in the text to be 0.6 V) can then be found from the gradient and the intercept of the graph respectively.

Now build the circuit given in *Figure 8.19(b)*. Use (preferably) a double-beam oscilloscope to observe the waveforms across capacitor C and at the base 1 terminal, across resistor R_4. Check that the waveforms conform to those mentioned in the text. Measure the frequency of the oscillation, using the oscilloscope, for the extreme settings of R_2, i.e. for a 5 kΩ and a 10 kΩ total resistance in that arm. Show that the frequency is, within experimental error, given by the expression

$$f = \frac{1}{CR \log_e 1/(1 - \eta)}$$

Transfer the oscilloscope to terminal B_2 and investigate the output waveform there. Is it a 'better' waveform for triggering purposes than the one you saw at the B_2 terminal? What criteria are you using?

Finally, remove R_1 and reduce R_2 gradually from its maximum resistance setting. At what point do the oscillations cease? Explain why this is so.

---

2 Describe the physical construction of a unijunction transistor and briefly explain its mode of operation.

3 The intrinsic stand-off ratio of a UJT depends upon the applied voltage V_{BB}. True or false?

4 The n-channel of a UJT is lightly doped and the p-type emitter is heavily doped. Why do you think this is done?

5 When holes are injected into the n-channel of a UJT from the emitter, why do these move towards the earthy end of the channel rather than the positive (B_2) end? What is happening to the electron component of the current at this time?

6 Suppose the circuit of Figure 8.4 to have the following component values: $R_1 = 100$ kΩ, $R_2 = 2.5$ kΩ, $C = 0.01$ μF. At what frequency will this circuit oscillate? (The same UJT is used.)

7 A diac in its on state can only be turned off by removing the applied voltage. True or false?

8 Describe the constructional form of a diac and explain its operation. To what uses can a diac be put?

9 Without referring to the text, draw the circuit symbols for (a) a UJT (b) a diac (c) a GTO (d) a triac.

10 What are the advantages of a gate turn-off thyristor over an ordinary thyristor?

9 Optoelectronics

Aims: At the end of this unit section you should be able to:
Describe the constructional details and operation of LED, LCD, filament and gas discharge displays.
Build simple circuits illustrating the use of seven-segment LED displays.
Describe the constructional details and operation of photoresistors, photodiodes and phototransistors.
Understand the principles of multiplexing displays and optoisolators.

There are a number of electronic devices which go under the general title of 'optoelectronics', and these in turn can be subdivided into two groups: light sources and light sensors. Light sources produce radiant energy; light sensors respond to radiant energy. Among light sources we have the light-emitting diode (LED), the liquid crystal display (LCD), and the hot filament and gas discharge displays. The first two of these are of most concern to us. Among light sensors are photoresistors, photodiodes and phototransistors. The two groups, sources and sensors, are brought together in fibre optics and optoisolator devices.

LIGHT-EMITTING DIODES

There are two main types of light source derived from semiconductor material: gallium arsenide (GaAs), which emits light in the near infrared region of the spectrum, typically at a wavelength of about 0.9 μm; and gallium arsenide phosphide (GaAsP), an alloy which absorbs the GaAs radiation and emits visible light in the red, yellow and green areas of the spectrum. The latter LEDs find wide application as cheap, low-power-consumption indicator lamps.

The principle of operation of these diodes is that in any forward-biased *p-n* junction, recombination of electrons and holes occurs in the vicinity of the junction when forward current flows. When such a combination takes place, the energy released by the electron returning to the valance band appears in silicon and germanium crystals in the form of heat, but in a gallium arsenide crystal as radiated energy, that is, a photon of light. The efficiency of this process is intrinsically very high; nearly all the released energy of the recombination appears as radiation. The actual intensity of the radiation is proportional to the rate of recombination and so to the magnitude of the diode current. The problem in manufacture is to extract as much of this radiation as possible, keeping in mind that the vital junction is buried between layers of *p-* and *n*-type semiconductor.

To enable light to escape from the device, special doping is necessary to prevent the bulk of the material from reabsorbing the radiation as it passes through and to ensure that internal reflection is reduced to a minimum. The protective casing also needs careful design to reduce surface reflection losses and to

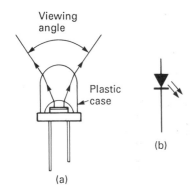

Viewing angle

Plastic case

(b)

(a)

Figure 9.1

avoid obscuring the junction area itself. A typical constructional form is illustrated in *Figure 9.1(a)*.

The protective dome is usually a glass or resin structure, coloured to suit the transmitted radiation and of refractive index intermediate between that of the *n*-type semiconductor (which is high) and the surrounding air. The dome also serves to magnify the light-emitting area of the diode junction. Although the actual light efficiency of red LEDs is rarely better than some 5 per cent, they are superior to yellow and green devices which have efficiencies of only 2–3 per cent. However, in relation to filament lamps they are much brighter in terms of power consumed because, unlike incandescent lamps, their output is entirely within a comparatively narrow part of the visible spectrum. Green also coincides with the maximum spectral response of the eye.

The viewing angle of an LED (see Figure 9.1(a)) is a function of the design of the packing rather than of the actual junction configuration. The casing may be domed or flat, may contain reflectors or diffusers, and so on. The TIL31, for example, has a domed casing which acts as a lens, but this restricts the viewing angle to about 40° overall. The TIL33, on the other hand, has a flat casing which allows a viewing angle of some 120° overall.

The electrical characteristics of light-emitting diodes are similar to those of conventional diodes except that the forward voltage drop of the LED is about 1.7–2.0 V against 0.7 V for the ordinary diode. The circuit symbol for the LED is shown in *Figure 9.1(b)*. The current required for an LED is typically within the range 5 mA to 30 mA, and in normal use a series resistor limits the current to a safe level.

The diodes come in a range of sizes, shapes and luminosities. There are the subminiatures with lens diameters of 2 mm, the miniatures at 3 mm, and the standards at 5 mm. These are the round, domed varieties. There are also the high-intensity types which have brightnesses up to ten times those of the standard range, and they are also available in sets having matched luminosities. This condition is necessary where a number of them are used in close proximity in some sort of display. Over and above all this collection, there are flashing varieties which have a built-in integrated circuit to switch the diode on and off at a rate of a few hertz, and bicoloured types with outputs of red/green or green/yellow light. Some applications of the basic LED, apart from a simple form of indicator lamp, will now be mentioned.

Bargraph modules   In addition to the common round types, LEDs are available in rectangular form, suitable for stacking either horizontally or vertically into arrays which can serve as level indicators, bargraphs, analogue meters and the like. Complete assemblies can be bought which contain a number of LEDs already stacked, together with the associated driving circuitry. Arrays are available in which the relationship between the input voltage and the number of LEDs illuminated follows either a linear or a logarithmic law.

The principle of the bargraph display is illustrated in *Figure 9.2*. Here we assume there are five LED indicators in the

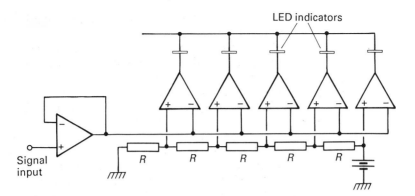

LED indicators

Signal input

Figure 9.2

display. The input signal is fed through a high-impedance buffer amplifier of unity gain. The output of this is applied to the array of LEDs by way of five comparators, one comparator to each LED. The comparators are each biased to a different voltage level by a resistance divider chain. As the input level increases, the LEDs light up progressively from left to right. For linear relationship, the divider chain is made up from equal-valued resistors; for logarithmic relationship, the values are appropriately graded. This kind of display is seen on most hifi amplifier equipment where it serves as a level indicator or as an output power level indicator. Dot patterns are often seen in place of the rectangular form of the array.

Seven-segment displays   These devices use LEDs in an arrangement which permits the digits 0 to 9 to be formed. They are familiar in most clock radios, a number of calculators, and test instruments such as digital counters. They are necessary, of course, to convert binary signals into decimal notation for the convenience of the observer. A seven-segment unit is shown in *Figure 9.3(a)*. Connections from the seven LEDs embedded in the base material are brought out to rear pins which can, if necessary, be plugged into an appropriate DIL holder or soldered directly into a printed circuit board. The common terminal may be connected either to all seven anodes of the LEDs (common anode) or to all

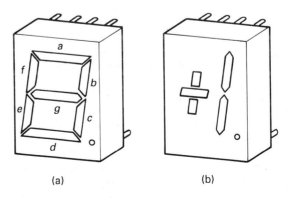

(a)   (b)

Figure 9.3

seven cathodes (common cathode), the choice depending upon the polarity of the applied signals. The other seven connections are lettered *a* to *g* as illustrated, and the individual LEDs may be lit in combinations which form the required digits. A format is also to be had which provides an indication of polarity in the form ±1; this is shown in *Figure 9.3(b)*. Decimal points are also usually included, and may be selected as right hand or left hand as required.

These displays are available in a range of character heights: 0.3, 0.43, 0.5, 0.56, 0.8 and 1.0 inch. The usual colours are red or green, though yellow can be obtained.

When a number of these displays have to be used, it is often inconvenient to build up an array from discrete units; it is then best to employ what are known as *multiplexed* assemblies, made up in two- or four-digit format in either common-anode or common-cathode versions. By end stacking such assemblies, an increased number of digits may be built up without the wiring getting too involved. We will return to the technique of multiplexing a little later on.

Hexadecimal displays   These displays are similar at first glance to the seven-segment types discussed above, but they are actually formed from a four by seven arrangements of dots. *Figure 9.4(a)* shows the pattern. Displays of the digits 0 to 9 are similar to those of the seven-segment variety, but the binary inputs for 10 to 15 are interpreted as the letters A to F. On seven-segment displays these inputs do not produce intelligible readouts. *Figure 9.4(b)* shows the resultant displays for the numeral 8 and the letters A and F.

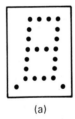   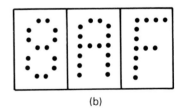

(a)   (b)

Figure 9.4

Alphanumeric displays or   In these devices, each digit consists of sixteen bar segments
starbursts   arranged in the pattern shown in *Figure 9.5*. They are not normally available in single units like the seven-segment displays. Instead they are sold in four-digit assemblies with built-in CMOS driving circuitry including decoder, multiplexer and memory. They are TTL compatible.

The digits 0 to 9, the complete alphabet A to Z, and a number of other symbols such as % + − ( ) & £, to a total of 64 presentations, can be produced on these displays.

Most optoelectronic LED display systems, when mounted in a finished piece of equipment, are placed behind an acrylic optical filter sheet which enhances the display image and reduces spurious reflections.

Figure 9.5

LIQUID CRYSTAL DISPLAYS

Unlike all other varieties of display systems, liquid crystals do not emit light but reflect incident light or transmit back light. These displays feature exclusively in watches, clocks and a great many pocket calculators, where their advantage of extremely low power consumption compared with LEDs makes them ideal for battery operation. They are not available in single units, but in $3\frac{1}{2}$, 4, $4\frac{1}{2}$, 6 and 8 digit, seven-segment format. The $3\frac{1}{2}$ and $4\frac{1}{2}$ digit types may seem curious if you haven't worked with them, but this form of designation simply implies that the first digit is restricted to the numeral 1; a $3\frac{1}{2}$ digit display, for example, is of the form shown in *Figure 9.6(a)*, while a full 4 digit display is shown at *Figure 9.6(b)*.

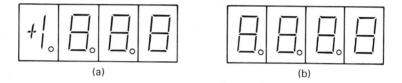

(a)   (b)

Figure 9.6

Liquid crystals are organic substances that are mesomorphic, being between the liquid and the solid state. A number of such substances are known, but the type used in displays can be electrically controlled. In the normal state, thin slices of this material appear to be transparent; this is because the crystals all face in the same direction (see *Figure 9.7(a)*). If a voltage is impressed across the slice, the crystals are disoriented and the transparency is affected.

For display purposes, a thin layer of liquid is sandwiched between two glass plates. A transparent electrically conductive film or back plane is put on the rear glass sheet and transparent sections of conductive film in the shape of the desired segments are coated on to the front plate. When the crystal is activated by the application of a voltage between the front film segment (or segments) and the back plane, the crystals within the liquid rearrange themselves so that the transmission of light from the back plane is interrupted. The incident light is scattered and absorbed within the liquid slice, and those portions of the liquid which are thus affected appear to be dark against a light background. *Figure 9.7(b)* shows the effect. It is not easy to see the segments or characters of an unenergized LCD, as it is with the LED displays, though by looking at the surface obliquely the pattern can be detected.

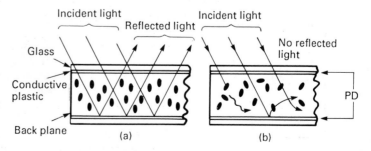

Figure 9.7

Liquid crystal displays cannot be used in the dark as they do not generate light. They are also slow in responding to signal changes; something of the order of 0.1 s is typical for a segment to change from light to dark. For this reason they are not amenable to multiplexing, and each segment of an array is driven separately. Actually, LCDs are not operated with direct voltages (which rapidly deteriorate the functioning of the crystal) but with low-frequency square-wave voltages of usually around 30 to 100 Hz. Types are available in which the segments appear light against a dark background.

FILAMENT AND GAS DISCHARGE DISPLAYS

Hot filament display modules have the usual seven-segment format but each segment consists of a thin wire filament. When energized, each filament glows in the same way as the filament in an ordinary light bulb, though not with such intensity. Typically each filament operates at 5 V and consumes about 10 mA of current. These devices are, therefore, wasteful of energy where a number have to be used, and battery operation is ruled out. Further, because of thermal inertia, their response is slow to signal changes. Like incandescent lamps, too, they have a limited life. Only one filament needs to burn out to render the complete unit useless.

Gas discharge displays consist of a segmented cold-cathode gas-filled indicator tube in a flat glass envelope, rather like a small valve. The segments of the cathode are arranged in the familiar format. When a voltage is impressed between the segments and a back-plate anode, the gas is ionized and the segments glow a bright orange colour. These displays are ideal for distant viewing.

A similar display tube, not so much seen these days, is the so-called Nixie tube. This is a multiple-cathode discharge tube, but instead of the usual seven segments there are ten distinct cathodes formed into the shape of the digits 0 to 9. Thus the required number is illuminated in its complete form and not as a selection of the appropriate segments.

Gas discharge tubes require an operating voltage of 150–170 volts, so they cannot be worked directly from the conventional TTL or CMOS decoders as LEDs can. These displays, as are the hot filament type, are often to be seen on the pumps at petrol stations.

SOME BASIC EXPERIMENTS

A few very basic experiments will soon make you familiar with the operation of seven-segment LED displays.

Get hold of a common-anode seven-segment module and, from the manufacturer's literature, make connections (a DIL holder is most convenient) to the segment and the common-anode pins. In series with each segment pin wire a safety resistor (220 Ω is suitable) and connect the anode pin to a positive 5 V supply. This arrangement is illustrated in *Figure 9.8*. Check on the individual segments by connecting, in turn, the free end of each of the resistors to the negative line. The *a* resistor should light the *a* segment of the display, and so on.

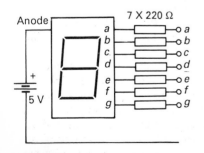

Figure 9.8

Check that the segments follow the lettering pattern given in Figure 9.3(a). Get this right or the rest of these experiments will not work properly.

Now use the appropriate groupings of the resistor connections to form correctly the ten digits 0 to 9 on the display. For example, by connecting resistors *a*, *b*, *c*, *d* and *g* together to the negative line, you should form the digit 3.

What we want the display to tell us next is the decimal equivalents of the binary numbers representing 0 to 9, that is, 0000 to 1001. For this we need a four-bit decoder which will accept the binary numbers and convert them into the arrangement of outputs feeding the seven segments of the display (just in the way we have already manually done) so that the right interpretation is made. The decoder we want is the 7447A (or the 74LS47 for lower power consumption), and the circuit this time is the one shown in *Figure 9.9*. The 7447A is a sixteen-pin DIL package and the pin numbers are shown in the figure. The input points, marked as A, B, C and D, accept the binary inputs. For example, if we connect terminals B and D together to the negative line (logic 0) and A and C together to the positive line (logic 1), the numeral 5 should be displayed.

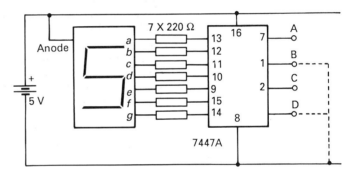

Figure 9.9

This is the decimal equivalent of the following binary input:

D C B A
0 1 0 1

Notice that D is the most significant bit and A the least. Also, if you leave A and C *disconnected*, the same display will be observed. This is because TTL circuits treat a floating input as high (logic 1), but a low input (logic 0) *must* go to the negative (or earth) line. Now work your way through the binary inputs 0000, 0001, 0010, ..., 1001 and verify that the decimal equivalents 0, 1, 2, ..., 9 appear on the display. Out of interest, see what happens when you insert the binary equivalents of the numbers 10 to 15. Why cannot you go beyond 15?

We can now make up an elementary counter. For this we need a decade counter package, which is found in the 7490 or the 74LS90. This integrated circuit receives input pulses and counts them in powers of 10. The outputs are at the pins marked A, B, C and D, and these give the count (from 0 to 9) in binary form. By wiring these pins to the corresponding pins on the 7447A, the count is displayed in decimal notation by the display. Feed a very slow input pulse train from a suitable oscillator (no faster than 1 Hz preferably) into the 7490, and the count on the display should run from 0 to 9 and then repeat. You should now be able to deduce for yourself how, by using more of the integrated circuits and additional displays, you can count input pulses up to, say, 999. You will find more details of this in Section 11 of *Electronics 2*.

MULTIPLEXING   Many integrated circuits which are concerned with counting, frequency measurement and the like, when connected to arrays of display modules, have their own multiplexing circuitry built in; the problems associated with the individual wiring of the displays are thus eliminated. As an illustration of what multiplexing means, we will suppose we have a counter made up from six circuits of the form shown in *Figure 9.10*. This counter will have six seven-segment displays, each driven by a 7447A decoder, with each of these fed from 7490 decade counters. The total count would be 999 999.

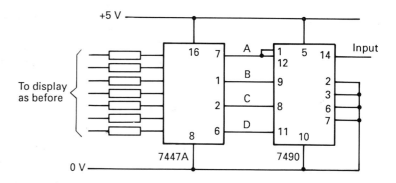

Figure 9.10

The current drawn by this assembly is surprising. Assuming 15 mA for each display segment, the total of 6 × 7 segments will draw 630 mA, and the six 7447As will take another, say, 200 mA. This is a total getting on for 1 A. The technique of multiplexing not only saves components but cuts the current consumption by a considerable fraction.

In this technique, the corresponding segments of all six display units are connected together as *Figure 9.11* shows. In multiple display assemblies, this form of connection is normally adopted by the manufacturers. Then each of the four outputs of a *single* 4774A go to the appropriate commoned segments. The anodes of the displays meanwhile are fed through individual driving transistors T_1, T_2 etc. Thus when the seven-segment code for a 6, say, appears at the output of the 7447A, it is bused to *all* the displays. However, if it is arranged that only one of the driving transistors in the anode leads is turned on at that moment, only *one* of the displays will show this 6.

The actual process is difficult to explain in a few words. The output of a switching circuit selects which decade counter's output is routed to the 7447A. The same switching circuit also

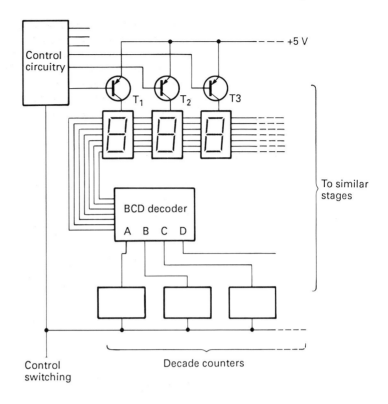

Figure 9.11

controls which driver transistor is turned on and therefore which display is activated. The switching process is synchronized so that, for instance, when the most significant digit data is being sent to the displays, the driver transistor for the most significant digit on the display is turned on. If this switching is done rapidly enough, typically at 50 to 200 Hz, the eye sees the result as a continuous display.

To achieve the same brightness as an ordinary display would have, it is usual to double the segment currents in multiplexed systems. In spite of this, multiplexing cuts power consumption considerably, saves wiring and reduces the number of safety resistors to just seven.

---

1 One of the segments in a seven-segment LED display becomes disconnected. The figures 1, 3, 4, 5 and 7 are unaffected by this. Which segment is faulty?

2 A student inadvertently crosses over the *e* and *f* segments in a LED display. Does this distort any of the numbers formed on the display? If so which, and in what way are they affected?

LIGHT SENSORS A light sensor or photoconductor responds to radiant energy in the form of ultraviolet, visible, infrared, electron, X and gamma rays and nuclear particles. We shall be interested only in those which are affected by the visible and infrared parts of the spectrum. Although photoconductivity was first discovered in about 1870, most of the progress made in understanding the phenomenon has followed hard upon the development of the transistor. Up to the early 1950s, photoconducting cells were made only of selenium or copper oxide; today commercial cells use germanium, silicon, cadmium sulphide, cadmium selenide or lead sulphide. *Figure 9.12* shows the intrinsic conductivity (undoped material) of these commonly used photoconductive materials as a function of the wavelength of the incident radiation.

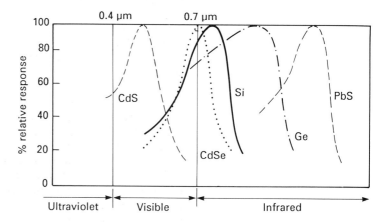

Figure 9.12

Those light sensors in which we are interested come under the general heading of photoelectric devices, and these can be group into the categories of photoresistors (light-dependent resistors), photodiodes and phototransistors.

THE PHOTORESISTOR

The photoresistor is a passive light-sensitive device, the resistance of which decreases as the incident light intensity increases. Its action is based on the fact that external energy provided in the form of light can act to remove electrons from their parent atoms, so creating mobile charge carriers which reduce the effective resistance of the material. *Figure 9.13(a)* shows the constructional form of a photoconductive cell, with *Figure 9.13(b)* illustrating the principle of operation.

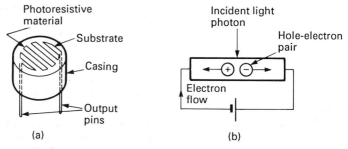

Figure 9.13

Photoresistors do not operate with *p-n* junctions and do not generate a voltage when illuminated. The most commonly used materials are cadmium sulphide and cadmium selenide. The former has a spectral response close to that of the human eye, while the latter peaks in the near infrared. When light falls on the photoconductive surface, the radiant energy contained in the light photons breaks some of the covalent bonds, causing the formation of hole-electron pairs. The application of an external voltage then causes a current to flow in the circuit whose amplitude is a function of the generated carrier density and hence of the intensity of the incident light. These cells are not polarized and may be connected either way round.

The semiconductor material is deposited on an insulating substrate in a zigzag pattern by a masking process as depicted in Figure 9.13(*a*). The whole is then enclosed in a glass or plastic envelope having a transparent window area over the sensitive surface. In some cells the window is coated with a thin film of gold, which serves as an electrostatic screen. Leads are brought out from each end of the semiconductor track through the base of the casing; in some devices these leads are made from a special material which minimizes thermocouple offset voltages appearing at the junction of lead and track.

The steady-state resistance of a photoresistor in the absence of illumination is called the *dark resistance*. The dark resistance of these cells can be as high as 10 MΩ, which falls to a few hundred ohms in daylight (the *lit resistance*). The lit resistance of a cell is determined not only by the resistivity of the photoconductive material but also by the geometry of the deposited track.

An important characteristic of these cells is the speed of response, which is a measure of the time required for the device to change resistance in response to a change in light intensity. Photoresistors have *response times* which are different for a given change in illumination from dark to light (the resistance fall time) than from light to dark (the resistance rise time). When light is applied to an unlit photoresistor, valence electrons are immediately excited to conduction. When the light is removed, electrons are no longer excited to the conduction band, but conduction does not cease immediately. The actual mechanism is complex and involves what is known as electron trapping, but conduction in a photoresistor whose illumination is abruptly removed continues until all the free electrons have dropped back into the valence band. Taking the response time as being between the 10 and 90 per cent levels, typical figures for photoresistor cells are 0.1 s for the rise and 0.35 s for the fall. They are not suitable, therefore, for situations in which the light intensity is changing rapidly, as in chopper amplifiers.

Figure 9.14 shows a simple circuit illustrating the use of a photoresistor cell to operate a relay. When light falls on the cell, its resistance is low and the base current is high. The transistor is switched on and the relay is energized. When the light is interrupted, the cell resistance goes high, the base and collector currents fall, and the relay is de-energized. The actual light level at which the switching occurs is set by adjustment of potentiometer R_3. The diode across the relay coil prevents the

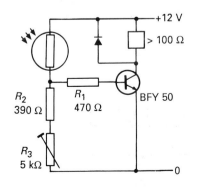

Figure 9.14

back EMF generated when the coil is switched off from damaging the transistor. Use an ORP12 photoresistor and make up this circuit for yourself. If you haven't a relay, use a low-current bulb.

> 3 What would be the effect of reversing the positions of the photoresistor and the two resistors R_2 and R_3 in this circuit?

PHOTODIODES

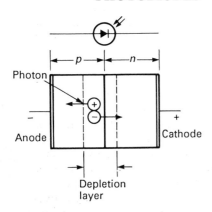

Figure 9.15

These, together with phototransistors, come into the class of photoelectric devices known as active devices. The construction of a photodiode is similar to a conventional diode except that the casing of the photodiode has a transparent area so that light can fall on the junction. Silicon photodiodes consist of a thin wafer of *n*-type material into which a *p*-type layer is diffused to a depth of about 1 μm. This narrow wafer allows light to penetrate to the *p-n* junction with negligible loss in energy. *Figure 9.15* shows the junction with reverse bias applied and a consequent depletion layer established. This layer, being free of mobile carriers, acts as an insulator between anode and cathode, just as it does in a reverse-biased conventional diode. When a light photon of sufficient energy enters the depletion layer, it is absorbed and its energy is released in the form of the generation of a hole-electron pair. Under the influence of the applied field, these pairs are separated, the electron moving to the *n*-type cathode and the hole to the *p*-type anode. Hence a current, additional to the normal leakage current, flows in the circuit, even though the diode is reverse biased.

Figure 9.16 illustrates this effect in terms of the biased diode characteristic. At (a) the diode is shown before it is exposed to light; the characteristic is normal. At (b) the diode is shown under the effect of incident light. It is seen that the reverse current level has shifted downwards, but otherwise the curve is virtually unchanged. The significant difference is in the negative portion of the graph; here we see that, for a constant reverse bias, a change in the reverse current is obtained which is proportional to the light intensity.

For maximum sensitivity, the depletion layer should be wide. This can be done by increasing the reverse-bias voltage, though this increases the reverse saturation (leakage) current and is undesirable. A layer of pure (undoped) semiconductor material can, however, be introduced between the *p*- and *n*-type layers, and this leads to the so-called PIN diode (*p*-intrinsic-*n*). This has two effects: it reduces the junction capacitance, desirable from the point of view of fast response, and it increases the reverse breakdown voltage level. The latter is not so important in the present application, but the former is.

Silicon photodiodes are available with a high-gain integrated amplifier built into the same small package. These devices are of particular value where accurate measurements of low-level lighting are needed.

If a photodiode is used without reverse bias it is said to be operating in the *photovoltaic* mode. As such it is suited for use

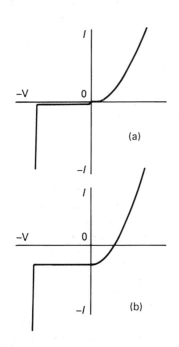

Figure 9.16

in optical instrumentation and camera control. In the photovoltaic mode, the electron-hole pairs formed when light falls on the depletion layer diffuse into the enclosing layers – holes into the *p*-type and electrons into the *n*-type material. These extra majority carriers on each side of the junction cause the terminal attached to the *p*-material (the anode) to become positive with respect to the terminal connected to the *n*-material (the cathode). An EMF is thus generated across the terminals which is proportional to the light intensity. This EMF varies typically within the range 50 to 500 mV on open circuit. *Figure 9.17* shows a suitable amplifier arrangement for a photodiode used in the voltaic mode. Using an FET operational amplifier, the value of feedback resistor R_f is chosen to give the required output in terms of the variation in diode current. Typically R_f will be large (470 kΩ to 1 MΩ), and the output is given closely by the product $I_D R_f$.

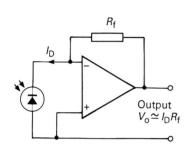

Figure 9.17

PHOTOTRANSISTORS

A phototransistor is essentially a photodiode with amplifier. It has the same basic construction as an ordinary transistor but is designed so that light can fall on to the base–collector junction. It used to be common practice in the early days of transistors to scrape the opaque paint covering off the plastic housing and turn normal transistors into phototransistors; the old OC71 was a firm candidate for this sort of treatment!

In the *n-p-n* phototransistor, the collector is biased positively with respect to the emitter so that the base–collector junction is reverse biased. So, when no light falls on the base, only the usual leakage current flows and this is negligibly small. When light falls on the base, however, more hole-electron pairs are created in the manner already described for the photodiode, and the current arising from these is amplified by normal transistor action. Essentially the base lead is redundant in a phototransistor, but it is often brought out of the casing so that additional collector current control over and above that generated by the light input is available.

> 4 The semiconductor material used in photoconductive cells is intrinsic; that is, it is undoped. Would doped material make the cell more sensitive?

OPTOISOLATORS

When a light source and a light sensor are brought together to act one upon the other, we have what is known as an optically coupled system. If the transmission path of the light from the source to the sensor is uninterrupted, the combination provides electrical isolation (an optoisolator) or a communications link (optical fibre communications). In the event of the transmission path being interrupted, the combination provides a means of counting (products passing by on a conveyor belt) or of measuring such parameters as speed, position and liquid level. We shall be interested only in optoisolators here, but the many possible circuit systems where light sources and sensors are used in combination must be appreciated.

Optoisolators are small units, usually packaged in six-, eight-or sixteen-pin DIL format, containing one or more infrared LEDs as the light source(s) in close proximity with one or more phototransistor(s) as the light sensor(s). The spectral response of each source and sensor pairing is closely matched, and they are separated by a medium that is transparent to the radiated frequency band. This medium, which may be air or some other such substance as glass or plastic, is an electrical insulator; hence there is complete *electrical* isolation between the input and output terminals of the packages.

Figure 9.18 shows the general theoretical circuit symbols for four optoisolators, each having an infrared LED source but with different sensors. At (a) the sensor is a phototransistor: at (b) the phototransistor is part of a Darlington configuration, and hence exhibits a considerably enhanced gain over the single transistor. At (c) the source is optically coupled to a thyristor which is triggered on reception of the light pulse from the source. And at (d) the isolator contains a triac which can be similarly fired. All these forms, and there are others, are designed primarily to separate possibly delicate low-voltage circuitry – as well as the human operator – from high-voltage areas in other parts of equipment. They also act as non-connecting interfaces in TTL and other logic systems. Voltage isolation up to 5 kV is provided by these units.

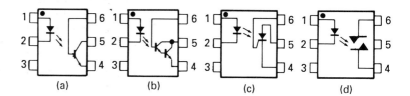

Figure 9.18

A typical application is seen in *Figure 9.19*. Here the optoisolator separates the trigger circuit of a triac from the mains supply. Although the LED is shown as operating from a 5 V supply by way of a hand-operated switch, the actual switching would most likely be derived from an accurate pulse generator.

High-speed optoisolators are now readily available which consist of an input LED optically coupled to a integrated detector made up from a photodiode, a high-gain linear

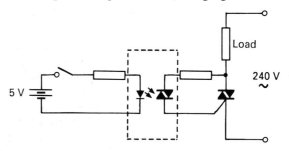

Figure 9.19

amplifier and an open-collector output transistor. This kind of isolator can be used in high-speed digital interfacing applications where common-mode signals must be rejected and ground loops eliminated. The problem with the ordinary phototransistor isolator is the lack of bandwidth, which stems from the fact that both the detection of the light energy and the amplification of the resulting photocurrent takes place in the one physical assembly. The large feedback capacitance between the base and the collector limits the signal bandwidth. Separating the photodiode from the amplifier reduces the feedback capacitance to a very small figure, and bandwidths up to some 25 MHz become possible.

In the next section we investigate fibre optics, where source and sensor form a unique communications link using optical transmission paths.

5 Is the radiation emitted from an LED when it is reverse biased or when it is forward biased?

6 An LED is to be operated at a current of 12 mA from a 9 V DC supply. What series resistor would you use?

7 An LED can be operated from an AC supply if a diode is connected in parallel with the LED, polarity opposed. Why is this necessary?

8 Describe the construction and principle of operation of a light-emitting diode, and give some practical applications of this device.

9 List the advantages and disadvantages of LED readout displays. Does an LCD overcome the disadvantages you have listed?

10 Explain the principle behind the operation of a liquid crystal display. Why are these displays used in battery-operated equipment rather than LEDs?

11 Why are LCDs not usually multiplexed?

12 Sketch the form of the numerals 0 to 9 you would observe if the *a* and *d* segment connections on a seven-segment module were reversed. Which numerals, if any, would remain unaffected?

13 Describe the construction and explain the principle of operation of (a) a photoconductive cell (b) a photodiode?

14 What is meant by the voltaic mode of operation of a photodiode?

15 A light source and a photoconductive cell are arranged so that objects passing between them can be counted. What special precautions might be needed to avoid false counting? The counter itself may be considered blameless.

16 A motorized trolley must follow a painted white track on a horizontal smooth surface. It has a sensor system made up from a lamp and two photodiode detectors which respond to light reflected back from the track. Suggest an arrangement whereby the trolley might be made to follow the track. Assume that only simple steering requires to be activated.

17 Describe an optoisolator module and give a few examples of its applications.

10 Fibre optics

Aims: At the end of this unit section you should be able to:
Appreciate the advantages of fibre optical transmission over other forms of communications link.
Understand the mechanics of fibre transmission.
Know the meaning of acceptance angle, critical angle and numerical aperture as applied to fibre optics.
Describe propagation modes and dispersion.
List the sources of transmission loss in optical systems.

Electrical signals can be transmitted in two basic ways: radio signals in space, and electrical signals along a form of metallic conductor. Light signals, which are of course simply extremely high-frequency electromagnetic waves, also form a means of communication, though we do not normally equate such signalling with electrical methods. Light transmission, however, has a number of advantages: it cannot be affected by either electrical or magnetic interference; two or more beams can cross without mutual interference or crosstalk; and, outside the limits of the beam, the signal cannot be 'overheard'. If instead of sending the light signal directly through space, where it will of necessity travel in a straight line, it is directed along a 'conducting' medium such as a flexible transparent fibre of glass, the signal remains secure and interference proof. It has the added advantage that it can travel around curves to its destination.

The introduction of fibre optics offers solutions to a number of the problems mentioned above which are associated with traditional wired transmission systems. It also has an advantage with regard to the transmission bandwidth. It ordinary coaxial cable the bandwidth available is proportional to length squared, while in fibre transmission it is directly proportional to length. Because of this, a greater number of different signals can be multiplexed on to a single fibre link for a given distance than is possible with coaxial cable.

Two types of fibre are in general use. A *step-index* fibre has a plastic-coated cylindrical core of silica or acrylic polymer of constant refractive index. In a *graded-index* fibre the density and hence the refractive index change throughout its cross-sectional area from a high-density centre to a low-density perimeter.

INTERNAL REFLECTION

At this point we go into the elementary physics of light refraction. When light, instead of reflecting from the surface of another medium, enters into that medium, the ray is directed along a different path in the second medium from that followed in the first. This is the process of optical refraction, and *Figure 10.1* illustrates refraction at a plane surface. At (a) it is assumed that medium B is optically denser than medium A; for example, medium A could be air and medium B glass. The light is passing from a less dense to a dense medium, and angle θ_1 (the angle of incidence) is greater than angle θ_2 (the angle of refraction).

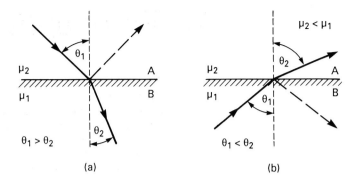

Figure 10.1

Diagram (b) shows light passing from a dense to a less dense medium. Here θ_1 is again the incident ray and θ_2 the refracted ray, and θ_1 is less than θ_2. Notice that the angles are measured relative to the line (the normal) drawn at 90° to the surface. Notice also, in both diagrams, that a small part of the incident ray is reflected at the interface.

The relationship between the angle of incidence θ_1 and the angle of refraction θ_2 is expressed in a law attributed to Willebord Snell:

$$\mu_1 \sin \theta_1 = \mu_2 \sin \theta_2$$

where μ_1 and μ_2 are constants, the *refractive indices* of the particular mediums concerned.

In fibre optics we are interested in the case of light being refracted from a dense to a less dense medium. Look at *Figure 10.2*. In this case, as you should have noticed from the previous discussion, the light is bent *away* from the normal as it passes

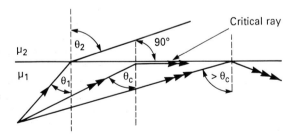

Figure 10.2

from the dense to the less dense medium. As the angle of incidence θ_1 in the denser medium is increased, the angle of refraction θ_2 in the less dense medium increases also, and as θ_2 is always greater than θ_1 a value of θ_1 will be reached where θ_2 becomes 90°. Since the light energy in the incident ray is divided between the refracted ray and the internally reflected ray, and since at a greater angle of incidence no refraction can take place, the whole of the incident light energy must pass into the reflected ray. For angles of incidence greater than the *critical angle* θ_c, the ray is totally internally reflected and none of it escapes into the less dense medium.

Since $\mu_1 \sin \theta_1 = \mu_2$, we can replace θ_1 with θ_c and θ_2 with 90°C. Then

$$\mu_1 \sin \theta_c = \mu_2 \sin 90°$$

But $\sin 90° = 1$, so

$$\sin \theta_c = \frac{\mu_2}{\mu_1}$$

If the less dense medium is air, for which the refractive index $\mu_2 = 1$, then

$$\sin \theta_c = 1/\mu_1$$

$$\mu_1 = \operatorname{cosec} \theta_c$$

NUMERICAL APERTURE

Suppose we wish to send a light signal along an optically transparent rod. It might appear that it would be sufficient to direct the light into one end of the rod (see *Figure 10.3*) and look for its arrival at the other end. While this system might possibly work after a fashion, there are a number of points to consider. What happens if the light does not enter the rod at right angles as shown? And what happens if the rod follows a curved path, as it must if such a transmission method is to have any practical significance?

Suppose we have a core of transparent material wrapped in some kind of cladding, as shown in *Figure 10.4*.

Figure 10.3

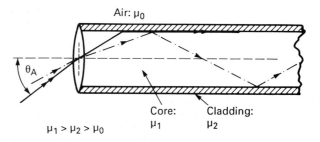

Figure 10.4

Consider light entering the core at a variety of angles, two of which are shown in the diagram. As light enters the core, refraction takes place at the incident face. However, once the light is inside, internal reflection from the core–cladding interface will only take place provided the incident angle of entry, θ_A, is *less* than some given value. At the critical point and beyond, the ray will not be internally reflected. θ_A is called the *acceptance* angle, and the sine of this angle is the numerical aperture. Hence

$$\text{numerical aperture} = \sin \theta_A$$

Then

$$\mu_0 \sin \theta_A = \mu_1 \sin (90 - \theta_c)$$

$$= \mu_1 \cos \theta_c$$

where θ_c is the critical angle for the core–cladding interface. But $\cos^2\theta = 1 - \sin^2\theta$. Therefore

$$\mu_0 \sin \theta_A = \mu_1 \sqrt{(1 - \sin^2\theta_c)}$$
$$= \mu_1 \sqrt{[1 - (\mu_2/\mu_1)^2]}$$
$$= \sqrt{(\mu_1^2 - \mu_2^2)}$$

But $\mu_0 = 1$; hence

$$\sin \theta_A = \sqrt{(\mu_1^2 - \mu_2^2)}$$

Example 1
The refractive index of a core material is 1.5 and of the cladding 1.42. What is the numerical aperture and acceptance angle of this core?

Here $\mu_1 = 1.5$ and $\mu_2 = 1.42$. Then

$$\sin \theta_A = \sqrt{(1.5^2 - 1.42^2)} = \sqrt{0.234} = 0.48$$

This is the numerical aperture. Then the acceptance angle is

$$\theta_A = \sin^{-1} 0.48 = 28.7°$$

Any rays entering the end face of the core at an angle greater than this (relative to the axial line of the core) would not be transmitted.

---

1 A certain fibre has a core of refractive index 1.46 and an acceptance angle of 30°. What is the refractive index of the cladding?

PROPAGATION MODES AND DISPERSION

Transmission along an optical fibre may be classified as monomodal, meridional or skew. *Monomodal* transmission takes place along a fibre that has such a small diameter (typically 5–10 μm) and numerical aperture that only one path, or mode, of light is possible. Meridional and skew transmission takes place along a multimode optical fibre (typical diameters being 50–100 μm) in which many light modes are possible. *Meridional* rays are those which pass through the axis of the fibre after each internal reflection, while *skew* rays never intersect the axis. Step-index fibre propagation is concerned with meridional rays.

In *Figure 10.5* the effect of input rays launched over a range of acceptance angles is shown. High-order modes are those rays which reflect a great number of times in their passage along the fibre, while low-order modes are those which make far fewer transitions. This variation in path lengths and hence in the time taken for different rays to propagate along the fibre, although all rays may be launched simultaneously from the transmitter, leads to a form of distortion known as *modal dispersion*.

This problem can be overcome by using graded-index fibre, whose refractive index changes gradually from a high value at the fibre centre to a lower value at the perimeter, as already

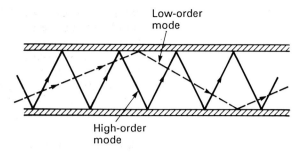

Figure 10.5

noted. Because the speed of light through a medium decreases as the refractive index is increased, higher-order modes, which spend less of their time in the central regions of the fibre, travel faster than lower-order modes. Hence the time differential between high- and low-order modes is much less for a given length of graded-index fibre than it is for step-index fibre. Generally, step-index fibre can be used in up to 10 km lengths, but this can be extended fivefold or more by the use of graded-index material in conjunction with a laser signal source which has a narrow spectrum of emission.

Modal dispersion is not the only form of dispersion in fibre optics. There is a form of distortion resulting from the fact that the velocity of propagation through a homogeneous medium is a function of wavelength. Consequently, the various wavelengths (or frequency components) of a signal may be launched simultaneously but not arrive at the receiver simultaneously. This form of dispersion is known as *material dispersion*. Again, such dispersion is only a problem on long-distance transmissions. An interesting fact is that for glass fibre links, the material dispersion figure is a minimum if the transmission wavelength is 1.3 μm, even when non-laser sources are used.

FIBRE OPTICAL LINKS A basic fibre optical link is shown in *Figure 10.6*. The signal is fed to an amplifier which modulates the emission from an LED. A special coupling feeds the light into the optical fibre link, and at the receiving end the process is reversed. The received signal acts on a short-wavelength sensor, usually an avalanche or PIN

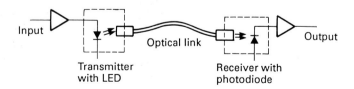

Figure 10.6

photodiode, which in turn drives an amplifier to provide the output. The transmitted signal has a frequency either in the visible red or the infrared part of the spectrum, typically within the range 550–900 μm. Both the LED and the photodiode are designed to cope with the frequency range concerned. The coupling units are constructed to give optimum light coupling from the transmitting LED to the fibre, and from the fibre to the receiving photodiode, and can be obtained as complete units to suit a particular fibre link. For laboratory experiments, visible light (red) is best as the operation of the link can then be followed by eye. In general, a loss of some 1–2 dB is experienced in the transmitting and receiving units, resulting from reflections from the fibre input face and manufacturing tolerances in the optical alignment of the acceptance angle. Coupling units are also available for joining lengths of fibre together, and these in turn introduce attenuation, again typically 1–2 dB.

TRANSMISSION LOSSES

As in any other communication medium, fibre optics has transmission losses. These come under three main classifications: material absorption and scattering, curvature losses, and coupling losses.

Material *absorption* results from the inevitable presence of molecular impurities within the fibre core which absorb certain wavelengths. Inherent impurities also cause the scattering of rays so that energy is dissipated by the unwanted breakup of the normal forward-moving light rays. *Scattering* also results from irregularities in the core–cladding interface so that perfect reflection from this interface is not achieved. *Figure 10.7* shows these effects. Fibres constituted from very pure glass are least likely to have serious losses from absorption and scattering, and are used over long-distance transmission paths in preference to polymer fibres.

Curvature can affect attentuation because, if the curvature at any point along the link is too great, part of the energy will be lost due to rays striking the core–cladding interface at angles less than the critical angle. Minimum bend radii are quoted by the manufacturers of optical fibres. In general, polymer fibres will work with tighter bends than will glass fibres, though some of this has to do with mechanical problems.

One of the few disadvantages of fibre transmission is the difficulty of coupling lengths of core together and of making connection to the transmitter and receiver units. As indicated, connectors are available but they tend to be expensive and they do introduce the bulk of the transmission loss in a particular system. The losses arise from misalignment of the joined faces, reflection at the faces, and the separation at the faces. A small gap is necessary to avoid scratching the faces when the coupling is made; such marks would make matters worse. It is best if coupling connectors are avoided altogether, or at best kept to an absolute minimum.

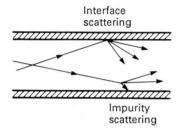

Interface scattering

Impurity scattering

Figure 10.7

Example 2
A polymer fibre is stated to have an attenuation of 1000 dB/km at a frequency of 660 nm. A 10 m length of this fibre is used to connect together a transmitter unit which introduces a loss of 2.5 dB and a receiver unit which introduces a loss of 1.5 dB. If the maximum power

injected into the transmitter at 660 nm is 15 µW, what power is delivered to the photodiode detector?

The cable has an attenuation of 1000 dB per km; hence for a 10 m length the attenuation will be 10 dB.
  The total power loss will be 10 + 2.5 + 1.5 = 14 dB. Hence

$$1.4 = \log P_i - \log P_o$$

$$\log P_o = \log P_i - 1.4$$

But $P_i = 15$ µW; hence $P_o = 0.56$ µW.

LABORATORY EXPERIMENTS

A simple laboratory setup is shown in *Figure 10.8*, using visible light and a polymer optical fibre. Suitable transmitter LEDs and receiving photodiodes are available from a number of suppliers, together with a length of matching fibre. A 1 metre length is suitable for this kind of experiment. Only the relevant parts of the transmitter and receiver are shown, as the remaining amplification can be quite conventional. Some experimentation may be needed to the component values indicated, dependent upon the actual LED and photodiode used, but the basic circuitry should be satisfactory for the demonstration of the principles of fibre optics communication.

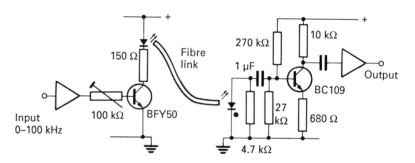

Figure 10.8

By feeding the input from a low-frequency generator, the effect of frequency variation can be studied, as can the effect on attenuation of bends in the fibre being made greater than the recommended minimum radius.

2 Name the advantages that fibre optics has over coaxial communication links.
3 Explain the following terms as applied to optical fibre communications: critical angle, acceptance angle, numerical aperture, meridional transmission and skew transmission.

4 The core and cladding of an optical fibre have refractive indices of 1.45 and 1.42 respectively. What is the critical angle for this cable? What is the acceptance angle and the numerical angle?

5 If the spectral sensitivity of the photodiode in Example 2 in the text is 0.45 A/W at 660 nm wavelength, what current will flow in this diode under the operating conditions given in the example?

6 Why would an ordinary incandescent lamp not be particularly well suited as a light source to transmit information over an optical link? In what circumstances might it be useful?

7 Explain the difference between step-index and graded-index optical fibres. *Figure 10.9* shows the path of light rays in these two types of fibre at (a) and (b) respectively. Explain the form that these rays take.

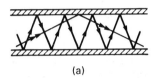 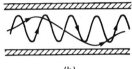

(a)   (b)

Figure 10.9

8 Transmission losses in fibres can be caused by (a) material absorption (b) material scattering (c) curvature radiation (d) scattering at the core–cladding interface. Briefly explain the mechanism of each of these causes.

9 Draw a diagram showing how the curvature of a fibre affects the critical angle condition for rays passing along the fibre.

11 Fault finding

Aims: At the end of this unit section you should be able to:
Understand the rudiments of fault-finding procedures.
Be familiar with the essential instruments required for basic fault location.

When you first look into an electronics system, however unsophisticated it may be in reality, you may feel inclined to say: 'This is not for me, I shall never understand it'. I can remember well my own first experience – a two-valve radio receiver given to me by a neighbour who built these things for himself in the days when there were weekly magazines all producing their own enthusiastic versions of the 'up-to-datest' in radio design. The word 'electronics' hadn't been coined then; it was all 'wireless' or 'radio', and even 'radar' lay some years ahead. None of us thought that an assemblage of valves, resistors, inductors and capacitors (condensers, as they were then) would ever do anything other than provide us with audible entertainment.

The two-valve receiver the neighbour gave me was assembled on a piece of heavily varnished plywood, and had a polished ebonite panel which carried a circular tuning dial, marked in degrees, and what was known as a reaction control. Inside, two massive valves stood proudly erect in two equally massive patently non-microphonic valveholders. There was an inductor wound with heavy-gauge wire on a three-inch diameter ebonite former, a solidly cased intervalve transformer, and sundry resistors and capacitors, all held neatly in special holders secured to the underlying 'chassis' of plywood. This receiver operated from a large high-tension battery, 120 V as I remember, and a 2 V accumulator which had to be recharged every few days, and the output went to a pair of headphones. And of course it needed an aerial which stretched the length of the garden.

At that time, that piece of equipment was quite frightening to behold and contemplate. What I would feel if I was faced for the first time with some of the equipment available today is best left unrecorded. But all electronics can be understood and mastered by a process of logical thought – not that anyone ever remotely reaches the stage where he can say he knows it all – and so can the location of faults which either develop in a piece of equipment or are there from the moment of switch-on. Self-made gear is usually the easiest to service. You have assembled for yourself a piece of equipment (most likely from a published design) and it fails to work when you switch it on — a common enough scenario. The trouble nearly always arises from insufficient care and attention being given to the soldering, a wiring error, confusion over resistor colour codes, or diodes being put in backwards. A careful check over the assembly with these points in mind will almost always put things right. After all, you have assembled the circuit and presumably read what the designer has had to say about things to do as well as what not to do. But to be faced with a piece of equipment not of one's own building, most likely a complexity of tightly packed components on a printed circuit board, which *has* been working satisfactorily and now isn't; that is a different matter. It is very unlikely to be a fault in the soldering (though that is possible), but it certainly cannot be a case of wrong resistors, reversed diodes or wiring errors.

You have only two things – logically – to help you put matters right: what the equipment *should* be doing, and what it is *actually* doing. Which, in the ideal case, is nothing! For there is no worse fault than one which comes and goes as it fancies. So you have to look for symptoms and interpret what those symptoms imply. In the early days of radio, a damp fingertip placed on strategic points worked wonders. It still can if you know what you're doing. But instrumentation is essential in fault location, even if such instrumentation comes down to an inexpensive analogue multimeter. Let us look at a few of the basic instruments we need, keeping in mind that the only way to become familiar with them is to use them. Reading about them, and about fault finding, can at best be nothing more than a brief introduction. There is no substitute for hands-on practice.

MEASURING INSTRUMENTS   The most basic (and essential) instrument of all is the moving-coil *microammeter*. This is an analogue instrument which forms the development of multimeters. The moving coil is fundamentally an indicator of direct current, but is readily adaptable to the measurement of sinusoidal current by the incorporation of some kind of rectifer. Moving-coil meters are available (at a price) with full-scale deflections of 10 μA, but a general-purpose multimeter will have as its basic movement a meter with an FSD of 50–100 μA. This can then be shunted to provide current ranges up to perhaps 10 A.

By using series (multiplier) resistors, a voltmeter is derived from the basic microammeter. By using a 50 μA movement, the voltmeter (whatever the range) will have a resistance of 20 000 Ω/V, and for most service work this should be considered a minimum figure. Most multimeters nowadays, even the relatively inexpensive ones, have figures of this order and sometimes better.

However, in a great number of instances of voltage measurement on modern electronic circuits, even a high-resistance instrument of this kind will load the circuit excessively – that is, draw too much current – and so lead to errors in measurement or to circuit disturbance, and often to both. The use of a *digital voltmeter* is then necessary. These instruments are now commonplace and, in a lot of cases, little more expensive than analogue meters. They have a high and constant input resistance, typically 10 MΩ and above, and so have little disturbing effect on the circuit under test. Current is measured as the voltage drop across a known resistance through which the current is passed. They will also measure resistance and check diodes, and a number have a simple continuity testing facility where a bleeper sounds through connection to low-resistance circuits.

A *function generator*, though not an essential instrument, is nevertheless useful to have. This can produce sine-, triangular- and square-wave outputs over a frequency range extending up to at least 100 kHz. It is useful as a signal tracer, and the square-wave output can be used as logical clock pulses as well as the means of checking the phase response of amplifiers.

The *cathode-ray oscilloscope* (CRO) is one of the most expensive pieces of equipment, but is necessary if any serious experimental design work is being undertaken. It also performs as an efficient fault finder, particularly in digital systems. With the CRO, the waveforms of voltage and current can be delineated on the screen and a permanent record made either by photography or (much more down to earth) by tracing with a pencil on to tracing paper.

Most oscilloscopes are of the twin-beam variety. Two signal inputs can then be compared as regards phase relationship, pulse ratios, frequency and amplitude. There is usually a choice of sweep modes so that both high- and low-frequency signals can be accommodated. In the *alternate sweep* mode, one transit of the beam across the screen displays the input to channel 1; the next transit, suitably displaced on the screen by the vertical shift facility, displays the input to channel 2. The idea is illustrated in *Figure 11.1(a)*. For high-frequency inputs, when the selected sweep speeds are smaller than a millisecond or so, the eye does not detect the switching and both displays appear to be simultaneously present on the screen. When the frequency is low, however, the necessity of a slow sweep speed leads to the alternations in position being visible, and neither waveform is intelligible.

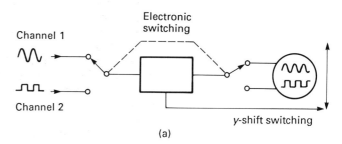

(a)

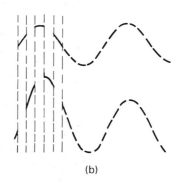

(b)

Figure 11.1

It is then necessary to use the *chopped sweep* mode, in which the beam switches very rapidly from one trace position to the other *during* each sweep transition across the screen. The beam consequently traces first a small segment of the channel 1 input, then a segment of the channel 2 input,then back to channel 1, and so on. *Figure 11.1(b)* shows the process, though in a much exaggerated form. Since the switching takes place at a very high rate (up to 100 kHz), both traces appear to be continuous and simultaneously present.

The input signal to a CRO is nearly always amplified before being applied to the deflecting plates of the tube. The frequency response (bandwidth) of this amplifier is stated to be so many megahertz, and for serious work 5 MHz should be considered a minimum requirement. It must be borne in mind that unless the response is adequate, the appearance of a pulsed waveform on the screen may be completely different from its actual form in the circuit under test. Hence a misleading interpretation may be put on its appearance and a fault suspected where, in fact, none exists. Knowing the limitations of instruments is often more important than knowing their virtues.

BASIC METHODS

In any fault-finding procedure it is vital to proceed in an orderly and logical manner. Jumping about the circuit, making measurements here, there and everywhere without thought, or touching things at random, achieves nothing. Neither does it follow that chasing injected signals from one end of a circuit to the other is the optimum way of solving things. The first thing is to make a precise note of what the fault symptoms are. Has the instrument been dropped, or has somebody poured a cup of tea into its bowels? Has it stopped abruptly, or has it given intermittent trouble for some time? Has anyone connected it to a voltage source outside its maximum input rating? From answers to questions like these, you can make a preliminary diagnosis of the fault from the symptoms and the possible causes, and from a knowledge of what the system should be doing under normal circumstances. This will often nail the fault down to a particular part of the system. That part can then be isolated and examined.

Suppose, as a simple example, an audio amplifier has gone 'dead'. When this happens, people tend to say: 'But it was working when I switched it off.' Of course it was; any piece of faulty equipment was working up to the time it broke down. But is the amplifier really dead, or just not giving an output at the loudspeaker? There *is* a difference; really dead would mean just that, nothing at all from the loudspeaker. But you might be getting a background hiss, so the amplifier wouldn't be completely dead, just failing to produce the wanted audio. That kind of distinction can make all the difference to the way the fault is approached.

Completely dead might mean nothing more than that the loudspeaker isn't plugged in. Check on such things always. Hours can be wasted searching for a 'fault' which results from a failure to plug something in – sometimes even the mains supply!

In the early days of radio I knew a man who wondered why his reception was dismal; he had connected the cord from the window blind to his receiver in mistake for the aerial wire which passed through the wall at the same point.

A totally dead piece of equipment more often than not means that the power supply has failed. Assuming that the mains power is present, checking on the low-voltage supply rails will quickly verify that the trouble is in the power unit itself. An external short-circuit can be eliminated by isolating the rest of the circuit. A possible power supply is shown in *Figure 11.2*.

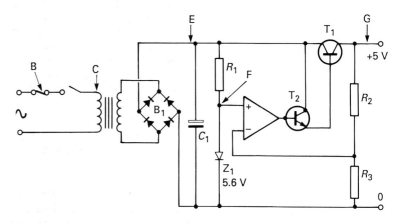

Figure 11.2

This is a simple regulated system providing a 5 V output at a current of possibly up to 1 A. A number of test points have been indicated on the diagram for the purpose of illustration.The test procedure could go rather as follows:

1 Checking at point G, using either an analogue or a digital voltmeter, should reveal an output of 5 V. If this voltage falls to zero or a very low value when the rest of the circuit is connected, the fault is a short-circuit somewhere within that other part and not in the power supply. If the output is zero, proceed as follows.

2 Check on the main fuse at point B. This can be done without removing the fuse by checking the mains input at point C, using an AC voltmeter set to 250–300 V. If this reads zero (and the mains is connected!), replace the fuse with one of identical rating. If this fuse immediately fails again, disconnect the mains supply and check for a short-circuit across point E, using an ohmmeter. If there is no apparent short-circuit (but allow for the fact that capacitor C_1 might charge up via the ohmmeter and so give that impression), isolate one end of the transformer secondary, replace the fuse and switch on. If the fuse again fails, the transformer is almost certainly faulty, although the mains switch might be going down to earth via its casing.

3 If the fuse is found to be intact at point B, check the transformer secondary output using the AC voltmeter. This should read (for this circuit) within the range 8–12 V RMS. If

there is no output and the mains supply is present at the primary terminals, the primary winding is probably open-circuit. The heavy-gauge secondary is very unlikely to fail in this way. Check for continuity using the ohmmeter, *after* disconnecting the mains input. If the secondary voltage is present, proceed as follows.

4 Check for the rectified output across point E, using the DC voltmeter. The voltage should be about 1.4 times the RMS value you measured across the winding. If this voltage is zero or very low, check the rectifier. Each diode can be tested in turn for low resistance in one direction and high resistance in the other. If there is an output across E, proceed as follows.

5 Measure the voltage across the zener reference at F. This should be 5.6 V. If it is low or zero, switch off and check the diode itself using the ohmmeter. If this is OK, check the feed resistor R_1. If this is OK, remove the IC and recheck the voltage across the zener. If the 5.6 V is restored, replace the IC with a new one.

6 If the voltage is correct at E and at F but there is no output at G, check the two resistors R_2 and R_3 feeding the inverting input of the IC. If these are normal, check finally on T_1 and T_2.

Now this may have sounded a bit of a rigmarole, and so in a sense it was. But what I have tried to show is that, even in an elementary application, the approach to fault finding should be logical; each new step is the direct consequence of the one before. If no steps are missed out, the fault, no matter how complicated the circuit, must eventually come to light.

There are often more tangible pointers to where a fault may lie in a circuit. A lot depends upon whether the circuit sits there quietly giving no outward signs of discontent, even though there is a fault, or whether there are manifest signs of disorder in the form of smoke, smell, fizzes or sparks. If any of the latter indications turn up, the rule is to switch off at high speed and search for the source of the heat or the smoke. This can be done by the cautious use of a forefinger; things which get hot very quickly under fault conditions are transistors, integrated circuits, resistors and electrolytic capacitors. Always make absolutely sure whenever you use a finger test in this way (quite a legitimate method of fault location) that all mains supply points are well out of the way.

Although such faults often appear in commercial equipment, the self-constructed instrument is usually the Parnassus of smoke and fire.

Electrolytic capacitors get hot because they have been connected up with reverse polarity. Apart from a gentle frying noise and sometimes a slight smell, they do not always give other signs of trouble until the end of the case blows off with a loud explosion and smothers the rest of the assembly with an unpleasant mess. Care is particularly needed in fitting the small tantalum bead type of capacitor, as the polarity markings are often very obscure.

The most likely cause of transistors getting hot is thermal runaway due to poor circuit design or inadequate heat sinking, or wrong connections. It is very easy to place a transistor, particularly those types with the three wire connections arranged in a triangular pattern, incorrectly into a printed circuit board. Simply reversing the collector and emitter leads rarely leads to overheating, though the performance isn't all it should be. The same applies to integrated circuits; the package can easily be plugged (worse still, soldered) backwards into its holder. The output of the IC might also be short-circuited to earth, or the output of one IC may be going high while the output of another is going low and there is an invalid (or accidental) connection between them. Solder bridging across the IC connections is a favourite troublemaker, as are solder splashes sticking between tracks. Any transistors or ICs which have been subjected to reverse supply voltages should be discarded, even though they might appear to be working afterwards.

Sometimes faults appear that are simply due to poor connections between the ICs and their socket contacts. Removing the IC and replacing it is often sufficient to cure this trouble. But be careful; unless you have a proper IC removal tool, use a small screwdriver to lift first one end of the IC and then the other, in turn, until it is free from its socket. It is the author's experience that to try to prise an IC from its holder using only fingers always results in buckled pins and perforated digits.

Resistors also tend to heat up when they are underrated for the job, or when there is a short-circuit between one part of the board print and another, often the result of poor soldering which bridges across the tracks or an odd strand of a badly trimmed wire is allowed to stray over to an adjacent track. Inspection of a printed board should always be done with a magnifying glass, preferably a watchmaker's glass that can be held in the eye. Breaks in the copper tracks can often be detected only in this way.

LOGIC CIRCUITS

For the testing of logic systems, a number of specialized though basically simple instruments have made their appearance.

It is not necessary to have anything more than a general understanding of the internal happenings in an integrated logical package to be able to use it or to test it. You can know what an AND gate does without the vaguest idea of how the internal circuitry performs the job. This of course applies to all ICs, but testing digital circuits calls for nothing more than an awareness and appreciation of a few basic ground rules concerning the power supplies, what is meant by high and low logical levels, and the input and output requirements of the various gates.

The most essential and elementary check to make on a piece of suspect digital equipment is to find out whether the power supply is turning out the right voltage for the system and that the power is available throughout the system. Every IC must have its supply pins energized. All that is needed for this is a voltmeter. As indicated, if there is no power and the system is dead, disconnect the supply from the system board and find out whether the loss of power is due to failure of the actual power unit or excessive loading due to a fault on the main circuit.

Always make sure you have the connection data for the ICs involved and that you know the pin numbers for the power input points. Working, say, with TTL, it is easy to get used to wiring pin 7 to earth (negative) and pin 14 (or 16) to positive V_s. However a common decoder, the 74141, has pin 12 to earth and pin 5 to positive, as have a number of other devices. Unless you are aware of this kind of variation, even a simple voltage check will give completely misleading results.

Most TTL and CMOS logic systems operate from a single supply source of +5 V, and in commercial designs this level is closely regulated. However, some tolerance is permissible and it is not unusual to find the supply somewhere within the range 4.5 to 5.5 V. The circuits operate within their logical requirements with supplies within this range, and no fault situation should be deduced from such supply tolerances in a practical system. Low voltage usually leads to longer time delays within the system, quite apart from the possibility of erratic operation, and this can be important where high-frequency counters are involved. Anything below 4 V or above 7 V (the TTL maximum) should be investigated.

In the case of CMOS logic, systems may be found operating from rails with voltages anywhere between 5 V and 15 V, though switching speeds are, like TTL, affected by low potentials.

Apart from such measurement of supply levels, quite a lot of logical testing can be carried out with a few basic instruments, of which three will now be briefly discussed.

THE LOGIC PROBE

For the quick determination of logic levels when exact values are not required, a logic probe is useful. This is a small hand-held instrument which derives its power supply from any convenient point within the equipment under investigation. In this way, the probe always responds correctly to the voltage level in force in any particular circuit system. A needle-like probe on the device is touched to various parts of the circuit where the logic level is to be checked; this is usually on the pins of the ICs themselves. The existing logic level at the point is then indicated on LED indicators. There are always at least two LEDs, and in many cases three. When the test point is at logic 0 (or less than 30 per cent of the supply voltage) a green LED lights; when the test point is at logic 1 (or greater than 75 per cent of the supply voltage) a red LED lights. A yellow LED is sometimes provided to indicate an indeterminate logic level. As an alternative, a two-colour (red/green) LED may be used in place of the single red type. In this way it will indicate red when the test point is high, alternate between red and green for a slow pulsed signal, and show yellow when the pulse rate is high. It will therefore give an indication of the presence of clock signals, for example, but some care must be exercised in the interpretation of the colour observed. Some probes use a seven-segment display which forms the letters H, L or P to indicate high, low or pulsing conditions respectively. Claims are made that this type of indication leads to unambiguous interpretations.

Whatever the circuit design, however, the logic probe is an inexpensive and useful piece of test gear, and no great problem to make up for yourself as an exercise in building – or even design. Think about it.

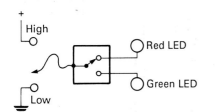

Figure 11.3

Example 1
Design a simple logic probe which will distinguish between high input levels (4–5 V) and low input levels (0–2 V), using two LEDs as indicators.

Figure 11.3 shows in a simple diagrammatic form what we have to achieve. What we want is some sort of electronic switch which will select the LEDs according as the input is effectively connected to the positive (high) rail or to the earth (zero) rail. The easiest way of doing this is to feed each LED from a transistor; each transistor is prebiased so that either a high or a low input at the probe point will switch on the appropriate LED.

For example, in *Figure 11.4(a)* the transistor is biased off by an appropriate selection of resistors R_1 and R_2. When the probe point (the base) is taken high (>4 V), the transistor is turned on and the red LED lights. Clearly, if the probe is taken low (<2 V) the transistor will remain off.

We cannot use the same circuit to switch on for low inputs, but if we replace the transistor with a complementary type (*p-n-p*) the problem is solved. In *Figure 11.4(b)* the transistor is again biased off by the right selection of resistors R_3 and R_4. When the probe input is high this time, the bias is increased and the transistor remains switched off. When the probe goes low, the bias is nullified and the green LED lights.

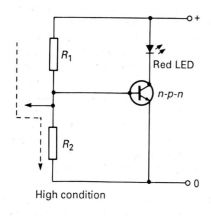

High condition

(a)

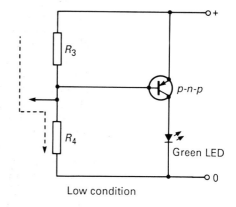

Low condition

(b)

Figure 11.4

A combined circuit is shown in *Figure 11.5*. Each potentiometer should be set separately so that the respective LEDs light for inputs >4 V (red) or <2 V (green). You are not likely to achieve this level of differentiation exactly, but there is no difficulty in making the probe distinguish between high and low logic levels. The diodes shown in broken lines are not essential, but they do serve to protect things if the probe is inadvertently touched on to a high-voltage point.

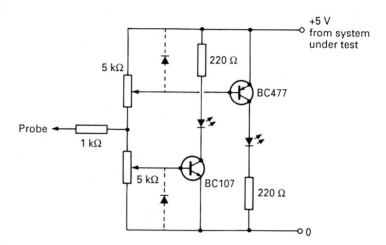

Figure 11.5

THE LOGIC PULSER

A logic pulser can be used in conjunction with a logic probe, but it can also be used on its own to 'break' into a circuit in order to alter the logical state at a particular point. A test on a shift register provides an example of this fault-location technique. As we know, data is shifted in a register from stage to stage on receipt of the falling edge of the clock pulse. So for a four-stage register, four clock pulses are necessary for a logical high at the input to be transferred to the output. Fault finding on such a register would normally involve a check that the clock pulses were present at the clock input of each stage (by oscilloscope or probe) and then following the Q output from stage to stage. This procedure is quite correct, but it does not necessarily lead to any conclusion about a possible fault if the data input stays static while the check is going on. So it becomes necessary to break into the circuit at the input and find out what happens when a logical low is loaded into all stages by connecting the first J input to earth, and then loading all stages with logical high by connecting the first J input to $+V_s$ via a suitable resistor.

To do this kind of break-in by unsoldering or making a break somewhere on the board track is not particularly enjoyable. A logic pulser, however, enables such tests to be made without disturbing anything. Essentially, the device produces an output pulse of relatively short duration (typically a few milliseconds) of either positive or negative polarity on operation of a control button. The current flowing into a short-circuit when the pulse is triggered is of the order of some 100–250 mA. In its quiescent (non-pulsed) condition the output is effectively dead, being of an impedance high enough to have no effect at all on the circuit to which it is connected. In operation, the pulser forces the existing logical condition at the test point to conform to the pulser output – not exactly a sledgehammer blow to a harmless nut, but enough of a sharp shock to the system to force things the way we want them to go. In other words, the effect is to 'toggle' the input to the gate under test, thus producing some sort of reaction regardless of the previous logical state of the input. The pulse, being of short duration, does not damage the previous stage; hence it is not necessary to break the circuit to isolate the connection between gates.

THE CURRENT TRACER

This is a device which provides a means of sensing and indicating the relative magnitudes of the currents present in circuit board tracks without having to break the track in order to insert a microammeter. In this way it resembles the logic pulser in that no circuit interruption is necessary.

Current tracers operate on one of two principles: either voltage dropped along a short length of printed track carrying a current is sensed by two probe contacts spaced a short distance apart (see *Figure 11.6(a)*; or a Hall-effect semiconductor chip senses the magnetic field surrounding a current-carrying conductor (*Figure 11.6(b)*). In both cases, the input is amplified and operates either an LED display or an analogue meter as an indication of the presence and relative magnitude of the circuit current. If, in moving along a current-carrying track, the current tracer indication suddenly reduces or disappears, a point of short-circuit has been detected.

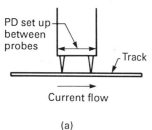

(a)

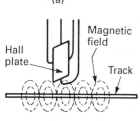

(b)

Figure 11.6

1 What are the advantage of a logic probe over a voltmeter for checking logic levels?
2 Describe a simple logic probe. How is it used?
3 What are the likely causes of (a) a hot resistor (b) a hot electrolytic capacitor (c) a hot IC? Consider cases of self-built circuits and commercial circuits.
4 What is the difference between alternate and chopped sweep modes in an oscilloscope? Which would you use if you were investigating two signals whose frequencies were greater than 100 kHz?
5 A voltmeter is stated to have a resistance of 500 Ω/V. What does this statement mean? Is this a good instrument? Where might it be most usefully employed?

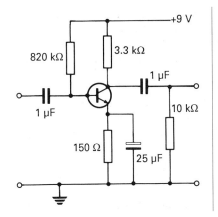

Figure 11.7

6 Under what circumstances does a digital voltmeter have an advantage over an analogue instrument? In what circumstances would the analogue meter score over the digital?

7 *Figure 11.7* shows a simple amplifier circuit. Which instruments would you use to show (a) the V_{cc} rail voltage (b) the base bias voltage (c) the signal at the input (d) the collector current? Do *not* disconnect any of the wiring.

8 *Figure 11.8* shows a simple logical gate system. Draw up a truth table for this circuit. You have only a logic probe at your disposal. Describe how you might test this circuit for proper operation. The 5 V supply is, of course, present and working.

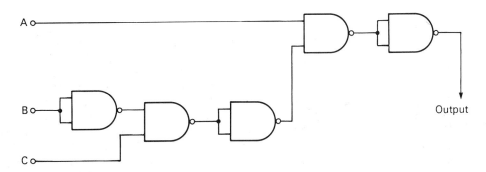

Figure 11.8

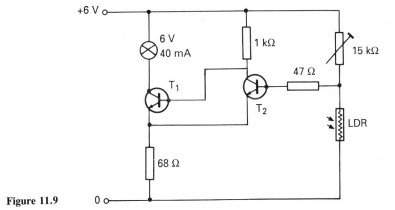

Figure 11.9

9 *Figure 11.9* shows the circuit of a simple light-triggered switch; when light falls on the light-dependent resistor LDR, the lamp in the collector of T₂ lights. It is found that the circuit fails to work even when the LDR is fully illuminated. Draw up a step-by-step report on the methods and instruments you would employ to trace the faulty component. Assume (a) that the circuit *has* been working correctly (b) that it has been built and has never worked properly.

Solutions to the problems

SECTION 1

1 40 dB, 32 dB, 14 dB, −6 dB; 100, 5, 4, 0.001.
2 (a) 0.316 W (b) 50 μW.
3 (a) 20 (b) 400 (c) 26 dB (d) 200 (e) 23 dB.
4 3 dB = 20 log (voltage ratio). For 3 dB reduction, voltage ratio = antilog −3/20 =0.707.
5 8.15 A.
6 40.
7 13.7 dB, 31 dB; 17.26 dB.
8 0.775 V.

SECTION 2

A_v / Without feedback / 10^4 / 5000 / 1000 / 100 / With feedback / 10 100 1000 10^4 10^5 f / Frequency (Hz)

1 9.93.
2 100.
3 300.
4 0.008.
5 β will be unity. Hence $A_v' = A_v/(1 + A_v)$, which is practically equal to 1. The voltage follower is such a circuit; see Section 5.
6 No. NFB reduces the noise output in exactly the same ratio as the gain. Thus for a given signal input, the S/N ratio at the output is unaffected.
7 The respective gains become 90.9, 98, 99, 90.9. The accompanying figure shows the new response curve.
8 10 dB.
9 β = 0.08. Bandwidth is 20.2 Hz to 1.237 MHz.
10 No; the base bias would be upset.
12 (a) True (b) false (c) false (d) false (e) true (f) true.
14 (a) 66.7 (b) infinity (c) 25. These solutions signify positive feedback, instability, and negative feedback respectively.
15 Positive feedback; β = 4×10^{-3}.
16 800.
17 90 to 91.7.
18 48.8; 47.6.
19 0.043.
20 29.7 dB.
21 R_i is doubled, R_o is halved.
22 30.7 dB; 0.03 times.
23 β = 4.9×10^{-2}; A_v = 2200 (use simultaneous equations).
24 49.7.

SECTION 3

1 No, because of the loss of gain.
2 Less. There are fewer carriers crossing the junction.
3 10^{-4} mW.

4 The S/N ratio will be unchanged at 40 dB since no more noise is added.

5 The signal is constant, so the noise must be reduced by 15 dB. This is a 31.6 ratio; therefore the bandwidth is 10 kHz/31.6 = 316 Hz.

7 Examples are (a) hum (b) hiss (c) ignition interference (d) lightning discharge.

8 90 μV.

9 26 dB.

10 0.5 μW.

11 See the text earlier, Example 2.

13 120 dB.

14 (a) 30 dB (b) 24 dB.

16 10 dB is a poor figure. A good figure would be 3–4 dB.

17 (a) 12 500 pW (b) 1250 pW (c) 11 250 pW. If the bandwidth is halved, the noise power will be halved.

18 (a) −84 dBm (b) −85.3 dB. 4.02.

SECTION 4

1 90°. The output voltage decreases.

2 L and C, or R and C.

4 Reversed coil connections, or the coil spacing too great.

5 18.4 kHz.

7 True.

9 Tuned collector; phase-shift oscillator.

10 90°.

12 The transistor input resistance is low, and this effectively reduces the Q of the tuned circuit. An FET would be better.

13 3.4 Hz.

14 (a) 9.025 kHz (b) about 16.

SECTION 5

1 −5 V.

2 −10.

3 (a) 4.7 MΩ (b) 4.65 MΩ.

6 As an impedance matching device.

9 100 kΩ.

10 −31.5 V.

SECTION 6

1 0 V.

4 50 kΩ, 25 kΩ, 12.5 kΩ.

SECTION 7

1 No.

2 I_H. The current at V_{BO} will be only slightly less than I_H, while the on voltage is practically the same as that at I_H (think about the thyristor characteristic). The current contribution to the $\propto$ values of the 'transistors' is therefore dominant.

3 It prevents reverse gate current. Yes, the circuit will work but the thyristor is at risk.

4 There is no phase shift on the gate to hold the thyristor off beyond 90°. It must fire at some point between 0° and 90°.

7 The thyristor will turn off if the latching level hasn't been reached when the gate current ceases.

10 (a) 4.8 A (b) 746 W.

SECTION 8

3 False.

4 So that under forward bias the emitter current is almost entirely composed of holes.

5 Holes are positive charge carriers and move towards the negative electrode, which is B_1. The electron component flows out of the emitter.

6 About 1.44 kHz.

7 True.

SECTION 9

1 Section *e*.

2 The numbers 2, 4, 5 and 9 will be affected.

3 The relay would be energized in the absence of light on the LDR.

4 No.

5 Forward biased.

6 Assuming a 2 V drop across the LED, 560 Ω would be about right in a preferred value.

12 Numeral 7 would be affected; also 6 and 9 if they were normally operating without 'tails'.

SECTION 10

1 1.145.

4 78°, 17°, 0.293.

5 0.25 μA.

SECTION 11

4 Chopped mode.

5 Not particularly good as it draws 2 mA of current for FSD. Only useful for measuring across relatively low-resistance circuits.

7 (a) Analogue voltmeter (b) digital voltmeter (c) electronic AC voltmeter or accurately calibrated CRO (d) digital voltmeter or good analogue voltmeter connected across the 3.3 kΩ collector load; then $I_c = V/3.3$ kΩ.

Index